RIGHT EVERY TIME

To Barbara, My Love

Think!

'To the thinking man life is a comedy
To the feeling man a tragedy'

Said Voltaire

This text invites you to think about it.
Think, and raise a smile.

RIGHT EVERY TIME

Using the Deming Approach

FRANK PRICE
Illustrations by Guy Smith

CRC Press
Taylor & Francis Group
Boca Raton London New York

CRC Press is an imprint of the
Taylor & Francis Group, an **informa** business
A TAYLOR & FRANCIS BOOK

CRC Press
Taylor & Francis Group
6000 Broken Sound Parkway NW, Suite 300
Boca Raton, FL 33487-2742

© 1990 by Frank Price
CRC Press is an imprint of Taylor & Francis Group, an Informa business

First issued in paperback 2019

No claim to original U.S. Government works

ISBN 13: 978-0-367-45602-3 (pbk)
ISBN 13: 978-0-8247-8328-0 (hbk)

Visit the Taylor & Francis Web site at
http://www.taylorandfrancis.com

and the CRC Press Web site at
http://www.crcpress.com

Library of Congress catalog number: 89-23251

Library of Congress Cataloging-in-Publication Data

Catalog record is available from the Library of Congress

CONTENTS

FIGURES

PREFACE

No writer can be all things to all people at all times, not even the committee of scribes who cobbled together the King James version of the Bible managed that. But it is reassuring for the writer to learn that the web of words he or she has spun has captured a fly of interest here or snared a moth of mutual meaningfulness there. It is a humbling experience when a reader, hitherto a stranger but now through the marvel of language a friend, tells you, 'Your book changed my life'. The balance of humility is redressed when another says, 'Oh, *you* are the person who wrote *Right First Time*', and pauses before adding, 'I *started* to read that.'

Then there is the reader who 'recognizes' the characters and the setting in what you have written. 'You *must* have worked in our place,' you are assured. 'You've *written* about it, how else could you know it?' Of course it is 'recognizable', it is all archetypes, a distillation of many people and many places. It is about nobody and nowhere in particular, so it is about everybody and everywhere.

All writing is a process of recycling the words written by previous writers, whose influence sometimes shows in what the French call 'l'homage' and others describe as 'plagiarism'. It is impossible for any writer to be rid of the influences which went to shape his or her style and I learned my craft from some of the greats: Gerald Kersh, the master of the short story; Helen Schuchmann who wrote that marvellous tome *A Course in Miracles*; the anonymous scholars whose ringing prose echoes in the Testaments . . . and many more. If I use some of the phrasing they have used before me it is less an act of thievery than a tribute to their excellence.

Words are the descriptive rafts on the rivers of experience; though one person can never 'share' another's experience it can be fun to be borne along on the raft and watch the river's banks unfolding. So please climb aboard, the journey is beginning, and may you enjoy riding this frail craft as much as its builder enjoyed putting it together.

Frank Price
Dyserth, Clwyd, September 1989.

1 THE LONG-NEGLECTED TOOLS OF QUALITY

> Man is a tool-using animal,
> Without tools he is nothing,
> With tools he is all.
>
> *Thomas Carlyle*

'With tools he is all'? Not quite. Many an amateur mechanic possesses all the tools of motor maintenance yet his car still sits immobilized because of some mechanical defect. Carlyle was only partly right. Ownership of the tools for the job is, of itself, never enough, to be effective tools have to be *used*, with skill and purpose.

The tools of Quality Management – the *statistical tools* used to transform process data into managerial control information – have been available to us in the West for long enough, yet until recently have generally hung rusting and unused. For how long? Since 1801, when the German mathematician J.K.F. Gauss published, in his *Theory of Number*, the knowledge which forms the basis of modern Statistical Process Control. This mathematical concept was taken up during the 1920s by an American physicist, Walter Shewhart, and turned into an elegant and powerful system for the controlling of process variables in engineering manufacture.

Thereafter it fell into disuse. The vast majority of Western manufacturing disdained the use of these statistical methods, preferring to depend instead on out-moded methods of inspection of product after the manufacturing event, when it is too late to make pro-active corrections. They preferred to produce scrap, and then complain about it. So its advocates – prophets without honour in their own land – turned to post-war Japan.

And Japan turned to them. Embracing this doctrine of frugality called statistical quality control. Adopting its methods. Adapting them. Improving them.

These tools of the mind, forged in Germany, tempered in America, honed to lethal sharpness in Japan, then found their way back to the hemisphere of their origin. In consequence the business landscapes of the Western world have undergone an upheaval – a Quality Revolution. It has shaken Western managerial thinking like some kind of earthquake of the intellect. This eruption of interest in total quality has sent its shock waves deep into the heartland of the manufacturing sector, its seismic rumbles down the slopes of the service sector, and its tremors along the business littoral into every commercial nook and cranny. With what kind of effects? Patchy, to say the least. Ranging from the cataclysmic to the inconsequential.

In some zones entire sectors have been laid waste. Once-proud companies, held together with nothing more than a crumbling cement of complacency, have collapsed into tumbled heaps of dereliction. Now they are no more than names in industrial archaeology.

Of the survivors a few, very few, have learned from the tremors of change. Their structures, stripped of vain adornment, are now buttressed with the girder-work of solid quality management, foundations strong enough to support their future. But many, it seems, look at the cracks in their organizational walls and wonder what to do next: a bit of plaster here, a prop or two there, and hope the insurance premiums are paid up to date.

It is as if Vesuvius had erupted and, having obliterated the decadent Pompeii, produced nothing more than a few bright Roman candles and a multitude of sputtering squibs.

For some reason too many of our enterprises still have not adopted the tools, techniques and philosophies of Statistical Process Control and Total Quality Management. They have the Means, they have the Material, but for some reason they lack the Mind. What is it that constrains them?

The constraints must lie within their *culture*.

Culture. There is a word more prone to misinterpretation than most. Even to utter it in the presence of some of our industrial chieftains is to meet the dismissive and contemptuous 'Culture? Don't talk to me about that kind of claptrap. We are here to work. I am the boss. I tell 'em what to do and they do it. We don't want any nonsense about "culture" in this outfit, thank you.'

That man has just made a *cultural* statement, although he is not aware of it. He has told you that the cultural climate in his organization is of the *power* style. He has also claimed that within this culture all power belongs to him; he is wrong, none of us can ever begin to be so omnipotent. He also denies the very existence of a thing called 'culture'.

You can no more avoid culture and its effects than you can avoid breathing that mixture of nitrogen, oxygen, trace gases and murky pollutants that we dignify by the description 'fresh air'. We are surrounded by culture, enveloped by it, drenched in it, permeated by it. Avoid it? The grave itself is no refuge from it. We are creatures of culture.

What do we mean by 'culture'? Is it as Matthew Arnold (in *Culture and Anarchy*) reckoned it to be – 'Culture being a pursuit of our total perfection by means of getting to know . . . the best which has been thought and said in the world'? Total perfection sounds much the same as zero defects, so this might be a fair definition of quality culture, but it is not a deal of use to us as a working specification.

Do we mean music of such haunting beauty that it speaks to us of the eternal within the temporal? Or the splendour of Dance? Or the new vision of a familiar reality bequeathed us by a long-dead painter? Or the majesty of language created by poets to enrich our imagination? Hardly. Culture these may be, but again we can hardly put them to work; they are too precious to be serviceable to our needs.

Do we then, when we refer to 'culture', mean it in a nationalistic sense – Mexican, or Swedish, or Russian, or Japanese . . .? Not really. This also affords us too constricting a concept.

Then what *do* we mean by 'culture'?

We mean the climate of belief, assumption, taboo, convention, supposition, folk wisdom, acceptance, understanding . . . all the nebulous things which together make up the psychological environment within our organizations; an environment which conditions attitudes, moderates behaviour, guides action. These are all to do with people's *minds*. If we are to make any changes to organizational performance we must address these cultural things; we must make fundamental change by changing minds because only 'mind' *is* fundamental. This is the principal purpose of implementing total quality management in the organization. Its primary purpose is to improve process and product quality, it *must* do this in order to acquire and retain credibility, it must pay for itself in the measurable terms of money earned by better quality practice. But this by itself is not enough, it is too superficial, too temporary. The secondary – yet principal – purpose of doing it is to change the culture in order that performance change will endure.

But every organization has its own unique culture – how are we to come to grips with such infinite variety in order that we may formulate some kind of working plan, a kind of road map, to help steer our actions in *any* organization regardless of how different it might be from all the other organizations with their own unique cultures?

What we need is some kind of management 'model', or 'theory', a standpoint from which to assess this entity called 'the organization'. We need something like . . .

Viewpoints

You walk through the city of Birmingham, you *see* the city. Then you drive through it in your car, you *see* the city, only now it looks a little different. You travel through Birmingham on the railway, again you *see* the city, only now it seems much more different. Finally you climb out of Birmingham Airport and you *see* the city unfolding beneath you, now it looks completely different. Four different viewpoints, which of these is the 'real' Birmingham? Obviously, they are all equally real, only now you know a lot more about the shape of Birmingham than you did after your first stroll through its streets, because each viewpoint afforded its own facet of that place we call 'the city of Birmingham'.

Now let us suppose there is a city called 'work'. It stands on a flat plain encircled by hills. From the summits of each of these hills we can gaze down upon the plain. Each hilltop provides its own unique view of the place called work. Each of these high vantage points is a 'managerial model', a concept, from which to examine the object of our enquiry. The more vantage points we have the wider our understanding of what we are looking at. This is what management models are for, this is why these sierras of the mind were erected, as standpoints to afford us different perspectives. Like their geographical equivalents these peaks have names: there is Mount McGregor, familiar to us for his Theories X and Y; there is Herzberg's Heights, whose view divides the rewards for working into hygiene and motivational factors; there are many more summits in this range of hills, many viewpoints.

Each of these viewpoints makes its own unique observations about the nature of work; not one of them is able to say everything, there cannot possibly be One Universal Model of Everything, but each says enough to be of service. But before we completely lose ourselves in a trackless wilderness of words which might be ceasing to mean much let us reconsider why we are bothering to contemplate these matters.

We are assuming that organizational performance is a direct expression of company or corporate culture.

Using the philosophies and techniques of Total Quality Management we aim to improve company performance.

We believe performance change will follow from cultural change.

We shall therefore try to consciously plan and bring about cultural change, by changing minds.

This is a daunting and demanding undertaking. We are speaking of *radical* change (radical meaning 'at the root', from the Latin 'radix'), by addressing the root causes where culture has its origins. To help guide us we need a management model of 'culture'. We have one. It is the *Harrison-Handy* model. Again, like any model this one does not – cannot – say all there is to be said about its subject, but it says enough to be of use to us. This is a high and appropriate vantage point from which to explore one aspect of the culture of work.

The Four peaks of Harrison and Handy

The purpose of a model is to simplify by uncluttering what would otherwise be too complex a picture to be understood. It is rather like a colour filter in a telescope, which takes out the dazzle in order that you might see more clearly. Somehow by looking at less of the total it reveals more of the essence.

Being engaged in that business described somewhere in the Old Testament as 'the labour of the foolish – the multiplying of words', I work from an office converted from one of the bedrooms of my house in North Wales. This stands on the eastern flank of the Vale of Clwyd, and a score or so miles to the westward the ramparts of the Snowdon range rear skyward to prop up their tent of cloud. The view through my window is constantly changing. It is not possible to see exactly the same view twice. Each of the windows' four vertical panes frames a particular peak on the Snowdonian horizon. At any time one particular peak stands higher than its fellows, then the curdled Atlantic sky sweeps inland on the prevailing wind to obscure it so that now another one seems tallest. A shift in the wind's direction and the canopy concealing another summit is drawn aside like a magician's cloak to reveal a higher pinnacle. The alterations to this rippling horizon are dependent exclusively on the capricious climate, they are beyond the observer's control. Even so, window-gazing can be a pleasant and sometimes rewarding way of passing time, in fact a consultant once remarked that I was conceptualizing. Dreaming a future.

I seemed to have spent ages gazing abstractedly through the window of my office at the horizon of smoky-blue hills to the west, looking for some unifying concept to encapsulate the cultural complexities of all the many organizations who were my clients. I needed some kind of mental structure which whilst embracing their diversity would filter out the dazzle and serve as a model to describe them all. Then when I met the Harrison–Handy model it dawned upon me that I had been staring at it all the time without knowing what it was. Those four peaks seen through my window were analogous to the four *styles* of culture in Harrison and Handy's model. Now they had new names. They are called Power, Role, Achieve (Task) and Support (People).

If you look at Figure 1.1 you will see these cultural summits depicted in an arbitrary order of comparative height. All that this model tells us is that in any organization any one of these styles predominates, and it is usually backed up by a second

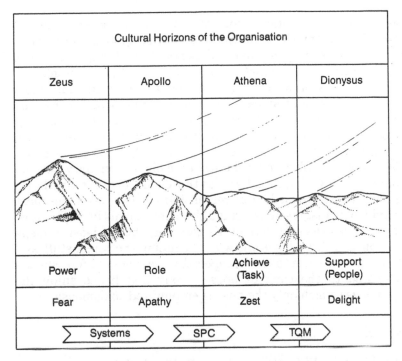

Cultural Horizons of the Organisation			
Zeus	Apollo	Athena	Dionysus
Power	Role	Achieve (Task)	Support (People)
Fear	Apathy	Zest	Delight

Systems ⟩ ⟩ SPC ⟩ ⟩ TQM ⟩

FIGURE 1.1 THE HARRISON-HANDY MODEL OF ORGANIZATION CULTURE (ADAPTED)

style. Just as each organization has a prevailing cultural climate which could be described fairly accurately (but not totally) by this model, so each of us – the organization's members – has a preferred style as described in this repertoire of four styles.

This is not to pass value judgements on any of these styles or on any person adopting any particular style, though it is hard not to do so. The model is morally neutral. There is no 'best' or 'worst' style, but there could well be styles which are appropriate or inappropriate to a company's situation or its declared business mission. These are matters of opinion. Let us look into the nature of the four styles:

POWER

This is the marionette-master style of managing. Always autocratic – '*I* am the boss', or even more '*I am* the company'. At its best it is paternalistic – 'I look after my workforce like a father' (yes, Dad, so by the definitions of your scheme of things your subordinates are perceived as dependent children); at its worst authoritarian (subordinates are seen as naughty children).

Companies, or even parts of companies, operating under this cultural style enjoy at least the benefit of having strength at the top. Or what looks like strength, but might in fact be something else. The boss-figure is the focal point; all orientation in the company is internally directed towards 'pleasing the boss'. When he is displeased he shouts. Like his patron-deity Zeus he smites with verbal thunderbolts those who incur his wrath. He seems to be aroused to anger all too easily – even the most trivial of transgressions is enough to unleash his biting tongue. Fear stalks the corridors of his company, but at least it keeps people in order. This is one way of running the whelk-stall, and some people seem to like it, especially the favoured few who are elevated to the rank of boss's crony and hence enjoy the privileges which only the boss is empowered to dispense.

ROLE

In this style the System is the system and all depends upon the system. It is a bureaucracy, feeding upon paper in order to produce yet more paper. Systems are necessary to the efficient prosecution of work; without systematization of our activities we would have to re-invent the rules every morning. But systems can take over, and are sometimes able to hold in thrall those whom they were designed to serve. They can somehow command the slavish obedience of the 'more than my job's worth' brigade of systems-followers who apparently never feel as secure as when they are restrained by the straitjacket of the system, as in the following episode from the annals of folly:

> One bright morning in April 1915 Hodges, John Henry, Sergeant, Lancashire Fusiliers, finally struggled through the submerged barbed wire and scrambled onto the Turkish beach at Gallipoli. Hundreds of his regimental colleagues did likewise, hundreds did not. The Turks, who had been waiting, made sure of that. By late afternoon he and his platoon had managed to cover the first fifty or sixty yards of their proposed trek to Constantinople and they

were running short of ammunition. He sent a runner back to the beachhead to fetch a case of cartridges from the ammo-dump by now brought ashore from the supply ships and piled onto the beach. The runner returned to the firing line empty-handed. 'Quarter bloke won't let me have any,' he said, 'not without an AB43.' An AB43 – Army Book 43 – is a stores requisition, one of the key documents of military bureaucracy. Hodges, a pre-war Tommy who knew the ropes, said he would fetch the ammunition himself. He belly-crawled his way through the gullies in the sand-dunes, arrived at the dump and requested a case of cartridges. The Quartermaster, another old sweat, also knew the ropes. 'Where's your AB43?' he asked as enemy machine gun fire cut yet another swathe through the growth of thyme that covered the dunes, tanging the air with its herbal fragrance to sweeten the acrid stink of cordite. 'Haven't got one,' Hodges confessed. The Quartermaster sucked his teeth and shook his head. 'No docket, no stores,' he stated with finality. 'All right then, pass me the pad and I'll make you one out,' Hodges offered reasonably. The man in charge of stores again shook his head, 'Sorry, mate, more than my job's worth. Got to be signed by an officer, you see, and you aren't an officer, are you?' Shells from the ships and shells from the shore batteries whistled and roared overhead in both directions as urgently as express trains. They had to shout to hear each other over the din. 'We haven't got any officers,' explained Hodges. 'They're all dead.' That Quartermaster was adamant. 'That's your tough luck, mate, no signature, no stores.' Hodges then committed a Court-Martialling offence. The butt of his service rifle blurred upwards in an arc that ended under the left corner of the Quartermaster's chin. Hodges then *stole* a box of ammunition. You can get shot for doing things like that, for bucking the system. Hodges didn't get shot, he was luckier, instead a Turk's bayonet pierced up through his jaw and out through his cheek, and one of his soldiers got him off by bayonetting the Turk more efficiently in this deadly game of tit for tat. After that he was commissioned in the field, and permitted to sign documents to draw stores officially from a bureaucratic system so beloved of officialdom.

Beware of systems, they can be dangerous idols to worship. In Role cultures the system is all. It's back to the bureaucracy of the Gallipoli beachhead all over again. In these Apollonian environments of carefully dotted i's and scrupulously crossed t's it is all 'please the system'. How about, for a change, *'please the customer'*?

ACHIEVE

If Power cultures are mainly to do with 'pleasing the boss', and Role organizations are concerned with 'pleasing the system', Achieve cultures are about 'pleasing the team'. These are brother-bonding organizations, typified by the task-force or commando group of interdependent problem solvers. This cultural setting provides an exciting and stimulating environment, making heavy demands on its members, who respond willingly to the challenge of the task. So it often predominates in the sunrise industries rather than in long-established smokestack organizations, though it can be found in the latter sorts, in such areas as Research and Development, as pockets of subculture within a different cultural style companywide.

This style of culture, whose patron deity is Athena the goddess of wisdom and war, seems to satisfy some deeply-held human aspirations, to appeal to a buried ancestral need to be part of a small élite of equals engaging in a common task. To a brotherhood, or to a hunting band. As in the hunter ethos this style is outward-looking, focusing on some external goal such as a market in which customers are perceived as targets, or some company goal external to the group – such as the tackling and completion of a project – which the task group is specifically set up to accomplish. This style is an internally motivated response to an external environmental challenge, as was the hunting group's response of old.

The questing spirit which is the ethos of the hunting band is still the driving force to progress and success. For this we should be grateful. Without it we would perish. This is where the ideas and methods of Total Quality Management find their profitable expression. Without this Achieve cultural style there can be no achievement. It is a culture we must cultivate, but not on its own.

SUPPORT

From the autocracy of the Power style, through the bureaucracy of Role and the meritocracy of Achieve, we come to the democracy of the Support culture. Where Achieve meant 'pleasing the team', this one is in essence about 'pleasing yourself'. This is the culture of the maverick, the person who though *in* the organization is not *of* it. As such it is just about meaningless even to conceive of an entire organization whose style is this. Small groupings of independent professionals perhaps, joining together in a loose federation for the sake of administrative convenience, a clinic of doctors, say, or a minor consultancy; but not an organization of any size. Its patron deity is the ungovernable Dionysus. This is the culture of the eminent professional who is absolute master of his craft; it cannot be faked. It is the 'achieve' carried to a level of almost-casual accomplishment. Every large organization needs at least one of these; but not too many, they tend to be unmanageable. Their personal freedom, which arises from their choice to be themselves because they are comfortable with themselves, is resented by some and coveted by others. It is resented most in Power Cultures.

But how are they managed? They are not; either they are led, or they are simply left to do their own thing; because what is good for the Dionysian is good for the company. These people lend themselves neither to the pursuit of lost causes nor to the achieving of trivial goals; they regard that most personal of things – their *time* – as being too finite and fleeting a commodity to be squandered in meaningless activities at the behest of others; they can be difficult subordinates. They can also be splendid subordinates, of immeasurable value to the leader who knows how to ride with a loose rein. Yet difficult and disruptive though they might be when handled with insensitivity, they are the essential agents of change. To this sort change is never seen as a threat, it is always a challenge. Being easily bored by the mundane and the humdrum they court change and seek to make change happen; they are often the leaders of it.

MIXED BREEDS

Every organization is a cultural mixed breed; none is a pure example of one particular style unsullied by any other. Sometimes it is a mongrel of all the four styles, living uneasily together in the same dog-house; more often it is a hybrid creature arising from the crossbreeding of two styles. The prevailing culture is always arrived at by accident, depending on what sort of people occupy what kind of positions of comparative influence within the organization. Only very rarely is a conscious attempt made to deliberately cultivate a certain chosen cultural climate.

Cultural style is important, too important to be left to chance. Every organization needs a good hunting band of 'Achieve' people to bring home the bacon, but without somebody to tend the campfires there would be chaos, so there has to be a necessary minimum of 'Role' culture bureaucrats to look after the chores which the hunters are too restless to see to. You are able to witness this tribal formation when you bring a big new project onstream. During the run-up of problem solving and troubleshooting, the 'hunting phase', the hunters are keyed up, the air is electric, the atmosphere exciting. Once the project is up and running the administrators move in; the prairie-busters have done their work so now it's the turn of the sweat-backs, the market gardeners who will tend the crops and keep the weeds down. The *great mistake* which the enterprise often makes at this stage is to say to those who pioneered it, 'You started it, so now it's yours, you run it.' *They are not fit to run anything* other than the hunt. They get bored, and when they get bored they make trouble. At this point they should abdicate and hand over to their legitimate successors, and having taken their ease and their rewards at the campfire for a while go out and do some more hunting. Where do they do this? The boss, whether a Dionysian or a Zeus, will find them something to do.

Culture is not only too important to be left to chance, it is too important to be ignored by the policy-makers and the mission-statement creators in the company, the people whose job is to get the best out of other people. Their awareness of 'culture' can be enlarged by reading Charles Handy's book *Gods of Management* (see Bibliography). This is no different from buying a complicated and expensive machine and reading the handbook before you begin to run it; yet for some strange reason it seems to be assumed that some people are 'naturally' fitted to run those infinitely complex structures we call 'organizations' by dismissing the handbooks as 'theoretical irrelevance'. Is there any wonder they sometimes clatter and bang along ineffectively, or seize up and judder to a stop?

Cultural Profile			
Zeus	Apollo	Athena	Dionysus
Power	Role	Achieve (Task)	Support (People)
Fear	Apathy	Zest	Delight

FIGURE 1.2   DRAW YOUR OWN MOUNTAINS

What About You?

What kind of cultural climate – Power, Role, Achieve or Support – prevails in your work organization? Which of the four peaks stands highest in the window and which hilltop backs it up? You can find this out by filling in the cultural profile shown in Figure 1.2.

The way to do this is to describe to a group of your colleagues the essential elements of the four cultural styles of the model. Then ask them to choose one style which they think most closely describes the climate which each of them believes surrounds them individually. Then simply plot the outcomes onto the profile and see which of the peaks is highest, and how it relates to the others. What you now have is a statement of collective *opinion*; just because it now appears on a kind of graph does not make it somehow 'scientific', it remains nothing more than opinion, but it is none the less useful for that!

Suppose you are a manager, and you invite half a dozen colleagues to play this little game with you. Each of you expresses his or her opinion about the prevailing cultural climate, and you dot these choices onto the scale in the appropriate pane of the profile window. Suppose five of you feel that you are immersed in an Achieve environment and two believe themselves to be living in a Support culture; the resulting mountain range now shows flatlands in the Power and Role panes and a peak rearing up to five levels in the Achieve, shadowing a secondary pinnacle two levels tall in the Support zone.

Now repeat the exercise, this time playing with a group of people from the shopfloor. Do not be surprised to discover that they see the Role Mountain as the highest, with a minor crag rearing to the leftward in the Power region; their flatlands are where your peaks are, in the two panes to the right hand side of the window. It's as if they are working for a different organization to the one employing your managerial colleagues and yourself.

They are. *Because they believe themselves to be.* And the job of 'management' is to change that belief. What is 'belief' all about? Let us consider the nature of belief and the universal need to believe. Every human society, from the smallest clan, through the tribe, to full nation, develops and uses its *myths*. There are three kinds of myth; *creation myths*, to explain the inexplicable fact of our very existence on this indifferent planet; *disabling myths*, to discourage us from doing those things which work to the detriment of society, by declaring certain acts, such as murder and incest, to be taboo; and *enabling myths*, which establish and reinforce those values and actions which society deems beneficial to its survival and propagation. The creation myth becomes embroidered into that complex system of belief and observance known as a *religion*.

A religion is 'Action indicating a belief in and desire to please a divine ruling power'.

The ruling power these days is the market and the customer, elevated to near-divine status. The 'religion' which believes in and wishes to please this latter-day god is Total Quality Management.

Quality, the New Religion

Quality, at one time a despised activity performed by despised people, but nowadays – thanks to Japan and Ford Motor

Company – a respected calling carried out by despised people, is becoming *a religion*. Not that there is anything wrong with religion, any more than there is anything wrong with mathematics or with motorbikes. The trouble with any religion is not so much the religion itself as the things done by its followers in its holy name. This is always an inherent danger with religion.

Nor is there anything new in the emergence of such secular religions. Ever since the advent of the Age of Reason a couple of centuries or so ago, when rational man deposed irrational God, the empty throne has served as a temporary lodging to a succession of fashionable deities. The trouble is that no sooner do people repudiate their belief in the One God than they bend in obeisance before any convenient substitute installed to fill the spiritual vacuum. Whether the surrogate belief be in dialectical materialism, scientology, the Puritan work ethic, Imperialism, capitalism, or whatever, its devotees, dazzled by the brightness of the latest enlightenment, are blinded to the fact that the feet at which they now choose to worship are still made of nothing more than common clay.

All these 'religions' – these objects of passing idolatry – spawn their priesthoods, their messiahs and prophets and gurus. Quality is no exception . . .

W. Edwards Deming, PhD, has been a consultant in statistical studies for over forty years, with experience worldwide in complex apparatus, industry, physical depreciation of utility plants, accounting, clinical and laboratory investigations, transportation and traffic. His clients include railways, telephone companies, carriers of motor freight, manufacturing companies, hospitals, legal firms, government agencies, and research organizations. His work in Japan created a revolution in quality and economic production and in new principles of administration. He was decorated in the name of the Japanese Emperor with the Second Order of the Sacred Treasure in recognition of his achievements in the Improvement of Quality and dependability of manufactured products through statistical methods. The Union of Japanese Scientists and Engineers instituted the annual Deming Prize, awarded to Japanese scholars for contributions in statistical theory or its application, and a medal, awarded to a Japanese company for advancement of precision and dependability of product. Dr Deming has received

many honours and medals, plus eight honorary doctorates. He
was elected to the National Academy of Engineers in 1983 and
inducted into the Science and Engineering Hall of Fame in 1986.

What is a messiah? *That* is a messiah. A messiah is 'an
anointed one; an expected liberator of an oppressed people or
country'. Deming certainly liberated the Japanese and Japan
from the oppressive thralldom of Third World starvation-
league poverty, into becoming the mega manufacturing nation
it is today.

What is a prophet? *That* is a prophet. A prophet is 'one who
speaks for God or any other deity, as the inspired revealer or
interpreter of His Will. An inspired teacher'. Deming surely
revealed the awesome power of statistical methodology as a key
concept of manufacturing and marketing strategy.

What is a guru? *That* is a guru. A guru is 'a Hindu spiritual
teacher, or head of a sect'. And the power of Deming's 'Methods
of Management of Productivity and Quality' appeals as
strongly to the 'spiritual' within us as it does to the 'logical'. As
much to the right cerebral hemisphere as to the left.

So, when we are dealing with Quality, we are dealing in
religious affairs. However, when we debate the Deming doctrine
and his Fourteen Points we must take the greatest care to avoid
the trap which every religion is prone to toppling into too
readily – that of embracing the Shadow instead of the Sub-
stance; of observing the Form and ignoring the Content; of
making it all Church and no Christ. This kind of thing can
happen too easily, and we are alerted to the danger by the
Director of Research of the British Deming Association, Henry
Neave who warns us about the *creative*, non-ritualistic nature of
Deming's Fourteen Points. '. . . They are vehicles for opening up
the mind to new thinking, to the possibility that there are
radically different ways of organising our businesses and working with people. . . . But there is great danger in just obeying the
words without first studying and developing deep understanding of why he is saying these things.'

But before we go on to examine Deming's doctrine, let us take
a short break for a bit of the other kind of culture, the literary
sort.

Life's but a walking shadow, a poor player

That struts and frets his hour upon the stage,
And then is heard no more; it is a *tale*
Told by an idiot, full of sound and fury,
Signifying nothing.

(W. Shakespeare, Macbeth)

A 'tale told by an idiot'? Why not?

The Idiot's Tale: 1
My name is Aaron Godman, and when I joined their company they said I was mad. At first I said I was not, in order
that the more I denied it the more they would believe it.
They did. There were more of them than there was of me,
their vote carried the day and confirmed me in my madness
in their estimation, and 'madness' depends mainly upon
who's put whom in the padded cell. Thus they willy-nilly
granted me the licence I sought, to say and do exactly what
I needed to. They bestowed upon me the jester's freedom.
So I conceded that being mad was OK because I was clearly
in the proper place – their madhouse. They called it a
'factory', and the MD (Most Demented?) called me 'an
enigma'. That shook me, did he know that one definition of
an enigma is 'a person both God and Man'? Had he somehow intuitively divined my true and dual nature?

Not an easy job, being a 'godman' – an apprentice
divinity. (Yes, I said 'apprentice'. Did you not realize that
even the lesser gods have to learn their trade? Why not in a
factory full of people? Where else in all the Worlds is
better?) No, not an easy job, but they – those who anointed
me in the time before your time began and told me 'thou art
ordained a priest for ever after the Order of Melchizedek' –
had never said it would be easy. Interesting, yes. Easy? No,
not while ever I clung to even the last tatty remnants of my
ego. Egotists bring war. My task was to bring peace.

So I became a Quality Manager in Bedlam plc. To bring
peace as a Quality Manager? As I said, not easy.

Why do I call it 'Bedlam'?

Bedlam – the Hospice of our Lady of Bethlehem – stood
two and a half centuries ago in London on the site now
occupied by the Imperial War Museum. An apt conjunction. In those days, if you were a gentleman of leisure, you
could fill the boredom of a rainy Saturday afternoon by
paying threepence to Bedlam's proprietor. This sum
bought you a ringside seat outside the cage within which
the poor inmates were confined, to be entertained by their
hilariously unpredictable antics. In those days insanity,
like bear-baiting and cockfighting, was a spectator sport. If
things were going a bit too slowly for your sophisticated

taste you could hire from Bedlam's proprietor, for a
modest sixpence, a pointed stick, with which to prod the
indolent insane into a more entertaining frenzy of gibber-
ing lunacy. Such was the rib-tickling nature of the times.

Times haven't really changed all that much. The only
difference is that these days you are in the cage with them,
and being paid for it. It's called 'management'.

You think this overstates the situation? Try this then . . .

British manufacturing loses an annual sum of between
£2,000 million and £20,000 million due to ineffective or
absent quality management.

Or try this . . .

Typically any Western manufacturing organization loses
anything between 15 and 40 per cent of sales revenue in
quality-related costs.

Does that sound like *sane* behaviour? Is it sane, or even
remotely sensible to 'lose' 30 per cent of bought-in raw
materials or components due to misguided quality
activity? This is management by madness. And my mission, as part of my apprenticeship, was to find out why and
put things right, as Quality Manager in an organization
dedicated to the crazy pursuit of generating waste. So I
joined the waste-makers, who called themselves 'manufacturers'. It was not going to be easy, nobody had ever said it
would be, but Deming's Doctrine helped, as we shall see.

Summary

The 'tools' of statistical methodology, developed in the West
and 'exported' to Japan over a generation ago, are now finding
favour with Western business. But their mere possession cannot, of itself, guarantee the achieving of quality. Like all tools,
the outcome depends so much on how skilfully they are used.
But all management, including that of quality, is a social
activity. So the effectiveness with which the tools are used is
dependent on the *culture* of the organization in which they are
employed. What do we mean by 'culture'?

The Harrison–Handy model, with its four cultural styles –
Power, Role, Achieve, Support – can be used to guide planned
cultural change in the organization, leading to improved quality
performance.

Quality – the new religion – and its most notable guru, Dr. W.
Edwards Deming. Aaron Godman's entry into Bedlam.

2 DEMING'S FIRST POINT

> 1 Create constancy of purpose toward improve-
> ment of product and service, with the aim to
> become competitive and to stay in business, and
> to provide jobs.

This is Deming's First Commandment, and there is no arguing
with it. 'Purpose equals Mission.'

So this is a *mission* statement. A mission statement is the
articulation of a *vision*. A vision is a dream-on-wheels which is
going somewhere. Whose dream? The dreamer's dream. Who's
the dreamer? Whoever in the organization happens to be
inspired. A leader.

Management is Administration.

Leadership is Inspiration.

Do you *like* being managed? Do you enjoy *managing* others? Or
do you prefer to be led, to follow a leader? And wouldn't you
rather *lead* than 'manage'? You will be invited to ponder this
question again and again throughout this text.

Management, as opposed to leadership, is more often than not
a delusion of control. It is a device employed by the fearful to
cast a grid of certitude over a future perceived as chaotic and
therefore threatening. It is an attempt to impose a reassuring
pattern of order onto a disorderly-seeming future by enforcing a
system of 'control' onto the present. It seeks to make the future
as predictable as the past, to make change unchanging.

How do *you* see the future? Perhaps you might like to try this
little experiment . . .

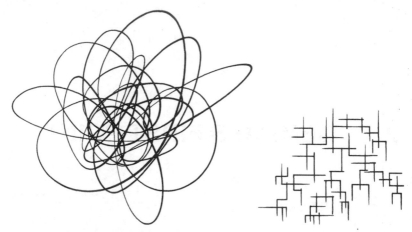

FIGURE 2.1   A TAKITI AND A GOLOOMA

Some years ago I saw, in a book, whose title and author I am unable to recall, a pair of pictures looking like those in Figure 2.1. One of which is called a 'takiti', the other is called a 'golooma'. Which is which? (Obvious really, isn't it?)

Now suppose that each of these represents an entire world, and you are required to decide which of these two worlds you would choose to live in; in which one would you feel most at home? Furthermore, which of these two worlds symbolizes 'management', and which 'leadership'? Each of these worlds has a characteristic climate; in which world is excellence most likely to flourish, do you think? In one of the worlds the games people play are of the win/lose kind, in which the winning and the losing bring any particular game to an end. These are finite games, played by finite players. In the other the purpose of the players is to keep the game going, without end. These are infinite games, played by those whom James Carse identifies as infinite players (see Bibliography).

You might be tempted to interpret one of these worlds as 'order' and the other as 'chaos'; be careful, things are not always as they seem. In fact, things are rarely as they seem.

There is nothing 'right' nor anything 'wrong' about which of these worlds you would choose as your abode, there is only your choice, and anything that the choosing says about you says it to

you alone; that is all, and that is enough. You have chosen your
world. You chose it long ago, a long time before you picked out
the takiti or the golooma.

Giving Away the Vision

So who, in the organization, dreams the dream, generates the
vision that there *has* to be a better way of doing things than the
way in which things are being done now? Anybody. Anybody
who is *dis*satisfied with the way things are; anybody who is
experiencing job *dis*satisfaction, not that cliché so close to the
heart of 'management developers' called job satisfaction; satisfaction does nothing more than smugly preen itself for being
satisfied. It is unwilling to face the prospect of change. So when
change happens, as it does constantly, it comes as a jolt to those
who are satisfied with the status quo. So the dissatisfied
dreamer of change envisions the way in which change should
take place. But unless the vision is shared it can never be
transmuted into a mission; unless it is *given away* it will remain
nothing more powerful or useful than a private fantasy, a
personal and impotent wish.

'Given away?' To 'give' is to lose, isn't it? Not always.
Suppose you are a dreamer (like Martin Luther King – 'I have a
dream'), you possess a *vision*. You 'give it away' by sharing it
with others. Do you now possess any *less* of it than you did
before you 'gave it away'? No, of course you do not, you still
possess all of it but you have *extended* it by sharing it with
others, it is impossible to lose it by 'giving it away'.

Golden Rule: You can give to nobody but yourself.

*So the role of the one who envisions a better future is to share
his vision, especially with the boss.* This is essential if the vision
is to motivate change. All the literature on the subject tells us
that the Chief Executive must demonstrate his 'commitment'
to Total Quality Management if it is to succeed in enhancing the
performance of his organization. What do we mean by 'commitment'?

COMMITMENT: A FARMYARD FABLE

The farmyard animals were gathered together in confer-
ence. The business on the agenda was to devise some small

token of appreciation for the farmer, as a way of expressing
the animals' approval of his excellent husbanding. The
horse, who loved a good conundrum, opened the proceed-
ings with the ancient Greek riddle, 'Is it better to be a
happy pig or an unhappy Socrates?' and stared interroga-
tively at the pig. The pig, who had heard this one before,
lifted his questing snout from his compulsive trough-
snuffling long enough to reply, 'Speaking as one who
knows, it is easy to be happy being a pig. Two banquets of
swill each day, an afternoon siesta on dry straw in an airy
sty, and the reassuring certainty that one is recognized as
being more equal than other animals – with this kind of
luxurious lifestyle it would demand considerable personal
effort to be anything less than happy. I am a pig, therefore
I am happy,' and he restored his snout to its proper place in
the cornucopia of the trough.

The goat, Speaker of the House, intervened. 'Order.
Order. This is irrelevant. The business of this house is not
to debate the mental health of piggishness, it is to decide
upon some token to mark the respect and affection we feel
for our esteemed benefactor, to wit, the farmer. Now, what
are we going to do?' The animals fell into a cogitative
silence disturbed only by the lip-smacking noises and
contented grunts of the foraging pig.

'I know what we'll do,' neighed the horse triumphantly.
'We'll make a nice breakfast for the farmer. How's that?'
'Good idea,' piped up the hen. 'How about making him
some bacon and egg?'

The pig's snout shot out of the trough. 'Hang on a
minute,' he squealed urgently, 'I'm not too happy about
that, *bacon* and egg? That's all right for you, you will be
merely *associated* with the breakfast; I, however, shall be
committed; no, sorry, I'm afraid it's just not on.'

The gathering erupted into an uproar of collective disapproval. 'Ooh, 'ark at 'im' jeered the ox, 'bloody typical,
isn't it, all take and no give when you're a pig.' 'Typically
swinish behaviour,' observed the sheep, who was taking an
Open University degree in sociology, 'rampant consumer-
ism epitomizing the least acceptable face of capitalism.'
The sheepdog, who was not taking a degree in anything
and who spent most of his time in famished indolence,
growled. 'Greedy ungrateful sod, that pig, always knew it,'
dropped his muzzle back onto the cushion of his front paws
and returned his attention to the rumbling emptiness of his
belly in the way farm dogs have ever done. 'Damned
spoilsport,' brayed the donkey, who had spent his forma-
tive years at a famous public school on whose hallowed
fields battles had been won, 'complete lack of commitment,

letting the side down.'

Under the overwhelming weight of collective peer group pressure the pig's resolution crumbled. 'OK then, I give in. Bacon and egg it shall be. I am committed.'

The farmer enjoyed his breakfast, especially the sizzling rashers of rash commitment.

But something odd happened. By a strange paradox the more the pig committed himself the greater and the happier he became; whereas the hen, being merely associated with the project, remained unchanged.

The horse rephrased the old Attic riddle, asking, 'Is it better to be a happy committed pig, or an unhappy associated hen?' Nobody bothered to answer him, because the answer was there for all to see.

COMMITMENT CASE STUDY 1: THE PROFLIGATE HEN

The boss of the company wished to improve his outfit's performance by using Total Quality Management as a vehicle for cultural change towards higher motivation and achievement. He had earmarked a large sum of money to fund the realization of this ambition. He had called in the services of a consultant whose job was to transmute this dream into reality. He and the consultant were discussing the project.

'So you will be running a series of seminars on quality-awareness education,' confirmed the boss, 'each seminar covering Statistical Process Control and Total Quality Management, each of one day's duration, to be delivered to a succession of groups of up to twenty people per group from different sectors of the organization. Sounds OK to me.'

'May I suggest that it would be beneficial if you yourself were to attend the first of these seminars', the consultant recommended, 'and that at each of the succeeding seminars there is a Boardroom representative attending as a delegate until every member of the Board has been exposed to the educational experience?'

'Oh no, no, no, no,' the boss shook his head, 'I'm afraid that sort of thing would not be considered appropriate, quite out of the question.'

'But you *are* committed to Quality?'

'Yes, of course I am, Quality is pre-eminent. In fact you will notice that I am committed to Quality to the tune of ... 'and here he mentioned the large sum of money budgeted to the quality improvement project.

'Yes, I see that you are prepared to spend money on the project, but are *you* really committed to its success?'

The boss was becoming a bit exasperated with this catechism. 'I've already told you, my job is to lay down the policy and allocate the necessary resources. This I have done. The actual project will be handled by my Quality Manager, that's what he's here for. The members of the Board are much too busy, and their time is much too valuable, to do as you suggest.'

The consultant deemed it wise to decline this lucrative assignment. Why?

COMMITMENT CASE STUDY 2: THE COMMITTED PIG

On the first of a series of one-day awareness seminars the Quality Management consultant was pleased to notice that the MD of his client company had honoured his promise to attend. At the end of eight hours' intensive training and education the consultant thanked the boss for having allocated a whole day to the seminar. 'Will there be a Board member at the next session?' he asked.

'Oh yes,' the boss assured him, 'on the next session you will have the Production Director sitting in.'

'Excellent, it demonstrates as nothing else can the strength of Board commitment to what we are trying to do.'

'I know,' nodded the boss, 'that's why I'll be sitting in at the next session as well as the Production Director. At the one after that you'll have the Financial Director – and me. The one after you'll have the Technical Director – and me; after that the R & D Director – and me *again* . . . after that . . . and me.'

This MD sat through a string of ten one-day sessions – nine of them repeats – over a three-month period.

'Don't you get *bored*,' enquired the consultant, 'listening to the same old stuff, replaying yet again the same old training exercises?'

He smiled. 'Bored? If all our people are able to see that the Board isn't bored with this quality thing, then nobody's bored by it. They all *know* that *we* are committed.'

'Arthur,' the consultant said admiringly, 'you're a hog for punishment, but by God I like your style.'

The company in question stamps hot metal into intricate shapes, in an environment of clanging smoke and flying sparks. Conventional wisdom in the trade decreed that this was too macho a business to lend itself to the fancy refinements of Statistical Process Control and Total Quality Management. Conventional wisdom was wrong. The MD knew it was wrong and he *acted* upon this knowledge and demonstrated his personal commitment for all to see by allocating ten of his very valuable days to the project. His company now uses quality management as a key element of its manufacturing and marketing strategy, and is the acknowledged leader in its field. *That* is commitment.

That is Leadership.

It was also one of the most rewarding assignments in the consultant's experience.

There is a lot to be said for being in the company of a happy committed pig.

So whoever is the dreamer in the organization who has an inspired vision of a better tomorrow, he must give it away, firstly to his boss. Then the rare earth of the vision will be wrought into the glowing metal of an inspiring mission, thus meeting Deming's First Commandment 'create constancy of purpose toward improvement . . .'

Improvement. To improve is to *change* for the better.

'Change' is an interesting subject. You can go on courses to learn 'the management of change'. Change is something which is being taken very seriously. 'Change is nowadays the only constant,' it is said. 'The pace of change is itself changing,' we are informed; 'it's getting quicker.'

On the other hand we hear 'people are afraid of change and resist it', and on the other hand 'a change is as good as a rest'. It's all a mite confusing, what are we to make of it all? How are we to respond to the pressure of an ever-changing environment?

Change

> In dim eclipse disastrous twilight sheds
> On half the nations, and with *fear of change*
> Perplexes monarchs.

So wrote John Milton in *Paradise Lost* more than three hundred years ago. The words are so relevant to our present situation they might have been written yesterday to describe today. Are we in the West in 'dim eclipse' now that the centre of gravity of manufacturing has shifted to the Far East? Are we unable to see our way forward any more in the long shadows of a 'disastrous twilight'? Time will tell. What about 'fear of change', does that 'perplex (our industrial) monarchs'? It certainly seems to have perplexed some of our industrial chieftains over the past few decades, judging by the outcome of their response to the challenge of change.

You could say that revolution is merely evolution accelerated to the point of being perceptible. Changes in climate and habitat, once proceeding at a pace so leisurely that they posed no obvious threat to survival, and so went ignored, now are whipped into a gallop so frenzied it is as bewildering as it is menacing. Survival now demands a speed of response fast enough to match the quickening rate of change, and the world of business is currently undergoing a process of binary fission – into the Quick and the Dead.

Four styles of response to this unfamiliar and unnerving state of affairs are possible:

1 DO NOTHING

This is the passive response. It has been practised by manufacturing organizations locked in a mindset of attitudes, assumptions and beliefs so rigidly glacial as to be unable to flex to the changing environment. Incapable of thinking the Unthinkable, too slow to join the Quick, they joined the Dead. You can witness their monuments, cluttering up the organizational landscape as thickly as the fossils of similarly failed species clutter up Kenya's Olduvai gorge. They are the derelict factories in our once-booming industrial heartlands, their roofs open to a leaden sky, silent save for the damp trickling down their walls like tears of sorrow for the stupidity of the leaders-without-leadership who betrayed their stewardship to a foreign invader.

This is the way of capitulation.

2 DO ANYTHING

This is the minimally active response. It could be described as a Philosophy of Reluctant Compliance. Its rationale goes something like – If we do nothing we shall surely perish, so clearly we must do something, but what? We don't really have much of an idea. OK then, we will analyse the situation. The sequence of steps in this shadow dance of death then goes – Analyse . . .

Analyse . . . Paralyse . . . Exhaustive analyses of every imaginable factor in the business equation – except the right ones – are undertaken. The malady besetting the enterprise is minutely diagnosed, but then, like a theatre of scalpel-toting but timorous surgeons surrounding an anaesthetized patient, each can name the affliction (wrongly) but none dares make the cut. Lingering death ensues.

This is management minus vision.

3 DO EVERYTHING

This is the maximally active response. It is the one embarked upon by organizations whose mindset is time-locked in the year of Gaius Petronius, AD 66, who observed:

> We trained hard, but it seemed that every time we were beginning to form up into teams, we would be re-organized. I was to learn later in life that we tend to meet any new situation by re-organizing, and a wonderful method it can be for creating the illusion of progress while producing confusion, inefficiency and demoralization.

So one popular response to the crisis of losing market share, and enduring negative returns on assets invested, is the shuffling of cards. Job cards are moved to different positions on that grid of Byzantine complexity known as 'The Organization Chart'. (It is also known as 'The Communication Chart' because this is what it is not.) Some cards are switched upward, others downward, others sideways, the scapegoat's is switched right out. Scapegoating is a phenomenon of organizational cleansing. It achieves nothing other than a collective sigh of satisfaction with the successful accomplishment of yet another sterile ritual. But every organization, like every church, needs its rituals.

These three styles of response to change have two characteristics in common. First, they are reflexive, in as much as they are neither more powerful nor more swift than the change which stimulates them. They are at best Newtonian 'equal and opposite' reactive responses. Secondly, they are isolationist, in that they behave as if the organization exists apart from its environment, so they seek the solutions to the problems of change, whose stimulus comes from outside the organization, by peering into the insides of the organization. But, above all, they lack creativity.

The fourth way has to be better.

It is. Some companies have found it, others have not. It is, of course, *Total Quality Management.*

Second-rate or Excellence?

With all the publicity that 'Quality' has enjoyed over the past few years you would think it nearly impossible that by now any company could remain unaware of it. Alas, you would be wrong. Even today there are still many organizations slumbering away in some kind of Rip Van Winkle limbo, who, according to a survey lately carried out by Dr Barrie Dale at UMIST, retain a resolute 'lack of knowledge and expertise' (as he tactfully euphemizes their condition of ignorance). Their potential for progress is also arrested by what he describes as 'lack of action from senior management', meaning either 'inertia' or 'apathy' or both; and 'poor understanding and awareness of what the whole business of quality is about' (more ignorance).

Now the main trouble with ignorance (ie with 'not knowing') is that the ignorant can never begin to appreciate the *extent* of their ignorance. This is the double-bind. You could therefore say that education is the process which renders visible the boundaries of ignorance, and once the boundaries are revealed education is able to push them back. The ignorant do not know this.

But even the words we use can be another trap: our mother tongue is not only a marvellous device for misleading each other during so-called 'communication', it is also an equally effective instrument for misleading ourselves. Take the case of a simple-seeming word like 'quality'. What do we *mean* by quality?

Here is where the confusion begins. Have a look at the 'graph' in Figure 2.2. What is the message in this plausible piece of nonsense? What is it telling you? ('Nonsense'? Yes, wait a minute and we shall see why it is nonsense.) It is *telling* you to settle for less than the best. To aim for the second-rate because you cannot afford the first-rate. It limits the aspirations of its believers, it legitimizes mediocrity. Its purpose, as part of the paraphernalia of the new religion of quality (and most religions, beginning as a celebration of life, are swiftly perverted by their

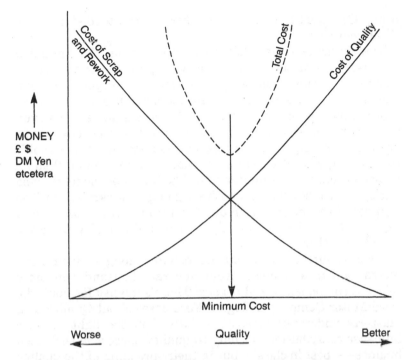

FIGURE 2.2   A 'MATHEMATICAL GRAPH' TO EXCUSE POOR PERFORMANCE

practitioners into a fixation with death), is to tell us that we are all sinners, beyond redemption, as good as dead, for whom excellence is for ever out of reach. So we shrug our spiritual shoulders, saying to ourselves 'Ah well, that's it then, nothing to be done about it,' and resign ourselves to something less than the excellence which is really attainable, as if mediocrity were divinely ordained.

This 'graph' is a lie! It is an amalgam of *three errors*; they are:

THE FIRST ERROR

It is based on the false assumption that 'grade' is synonymous with 'quality'. But these words *do not* mean the same thing at all. Consider, by way of illustration, the nature of the motor car.

Assume that 'Quality equals exact fitness for intended purpose'. That is only half a definition, but it will do for the time

being. OK, Quality equals exact fitness for intended purpose.
We accept this.

Now imagine – very simplistically – that the 'intended
purpose' of a motor car is to provide personal transport, to
enable you to move from one place to another. Forget, if you can,
any other 'purpose' served by a motor car. Put out of mind the
Motor Show, with a sleek new model on display and an even
sleeker and not so new model sprawled across the bonnet with
her hair dangling over her face and her eyelashes like rabbit
snares. Forget that aspect of it, that is nothing more than an
empty promise to the punter that if he buys one model it gives
him access to the other and both are going to bankrupt him. The
purpose of the motor car, in our definition, is *transport* – not a
'transport of delight' – merely a means of mobility from one
place to another.

Imagine further that you are driving along in your Ford
Sierra, a car assembled from several thousand randomly
selected components, and *it goes*. There's quality for you! The
Ford Motor Company's slogan 'Ford cares about Quality' is a
masterly understatement when viewed in the light of their
immense contribution not only to quality success in their own
business – 'best in class' – but in their spreading of the quality
gospel across the length and breadth of the world of manufacturing. They mean it! So there you are, motoring along, paying
the very greatest respect it is possible to pay to quality, which is
to say you are taking it completely for granted, and you pass a
Rolls Royce car, sitting on the hard shoulder with steam coming
from under its bonnet. (OK, unlikely, but *imagine* it.)

Which is the *quality* car? The one which is fulfilling its
'intended purpose', the Ford Sierra.

We all know very well that the Rolls Royce, stuck on the hard
shoulder or not, is worth many times the cost of the Sierra. That
is why people buy Rolls Royce cars; they say about you some of
the things you cannot, in all modesty, say about yourself. That
is why they are bought by successful gangsters, bookies, scrap-
men, pornographers, consultant gynaecologists – anybody who
is somehow vaguely ashamed of the way he gets his wealth. Oh,
they are also purchased by anybody who has an exquisite
appreciation of the supreme excellence of British automotive
engineering. (That's just in case *you* drive a Rolls.)

But the Rolls Royce is no more a 'quality' car than the Ford Sierra.

So the graph of 'quality' in Figure 2.2 is in fact a graph of 'grade' or 'class', or 'status'. But if we allow ourselves to believe it to be what it says it is, about 'quality', then we shall allow it to limit our goals to going for the second-rate. This is the sort of graph that Japanese businessmen might contemplate last thing every night before retiring, shaking their heads in wonderment that their Western competitors *still* believe in it, and falling into bed laughing at such folly. To repeat, 'Quality' is not the same as 'grade'.

THE SECOND ERROR

'You can't have quality and quantity' is one of the tenets of faith still firmly embedded in the mythology of Western manufacturing. (Mythology – that body of belief, unsupported by observation or experience, but steadfastly believed in none the less, which governs our actions because we always act as if what we believe to be true is really true. And before you scoff at myth, pause and reflect that some of the beliefs held sacred in the culture of *your* organization might just be a little bit mythical in their origin and nature. We shall be dealing with mythology, and *how to use it*, later on.) This myth is entangled with others ... 'The better the quality, the more it costs'. 'We are not in the Rolls Royce end of the game, we are in the Ford end' is, as we have already seen, a popular (and slanderous) excuse to make too much junk. 'Ours is a unique process'; they all are, so none is. 'It's the fault of the raw material'; is it really? Well, fancy that. 'Ours is a delinquent process'. Delinquent? Nothing like a bit of anthropomorphic projecting of human frailty onto an inanimate process for venting the spleen of frustration, you can always kick hell out of the car and heap abuse on it because it refuses to go. It will simply ignore your tantrums. All these cop-outs are enshrined in that pernicious depiction of the desirability of aiming for less than the best.

THE THIRD ERROR

This is the assumption on which the graph is built, that an organization actually *knows* its quality costs. Oh so very, very few know these. They see only the obvious tip of the iceberg of quality costs, usually the monthly sum of money paid out as credit notes to discontented customers as compensation for yet another load of rubbish shipped to the market. The other seven-eighths of costs are submerged, lost in the accounts, all those on-costs like paying for rework, transport charges for carrying substandard products back and forth to market and so on – costs arising from the second kind of work. (There are two kinds of work, work which adds value and work which adds cost, and the trick of good business is to maximize the former and minimize the latter.) These are lost in a mass of other data.

THE GOAL OF THE BUSINESS

So we shall resist the seductive siren-call of this graph which leaves us blameless in our pursuit of the second-rate, we shall *strive for excellence*.

Is this the goal of the business?

In his splendid and eminently readable book *The Goal*, Eliyahu Moshe Goldratt tells us: 'The *goal* of the manufacturing organisation is money.'

He is, of course, absolutely right. The manufacturing organization is brought into being to service a perceived market need. It can have no other purpose. No other *goal*. It is as if the *target* gives rise to the rifle, without which the rifle's existence would be meaningless. Of course, though the *goal* of the organization is *money*, it has as well obligations, duties, commitments, observances, such as 'Do not break the law of the land'; 'Do not pollute the environment'; 'Pay due and positive regard to local community needs'; 'Do not discriminate against minorities, like cigarette smokers or those of a different religious belief or skin pigmentation' . . . but only one goal.

Is this pursuit of money of itself a worthwhile goal? Certainly not for individuals; for individuals it is a recipe for a wasted life.

How about for organizations?

The story goes that when a market predator, on the acquisition trail, tried to take over the Pilkington Group, its boss informed Anthony Pilkington that 'the goal of business is to make money'. Mr Pilkington, head of the world's biggest glass-making organization, is reported to have replied 'Yes, but in our business we *strive for excellence*, and somehow money comes along.'

If striving for excellence is good enough for Pilkington plc it should be good enough for the rest of us. And who knows, money might come along almost as a by-product of our striving for excellence.

The Idiot's Tale: 2

Hello again, it's me, Aaron Godman, the Enigma. Remember me? I was going to tell you about that manufacturing company – 'Bedlam plc' – that I joined as Quality Manager. They were a big company with several factories, and their finger in many manufacturing pies, including the business of thermoplastic moulding. At that time the UK economy had just moved through the long transition phase from sellers' markets to buyers' markets. This is a crucial consideration in quality philosophy and practice. Quality is a function of supply. When demand outstrips supply, quality is simply irrelevant, the customer gladly accepts whatever is available, and makes the best of it. But as soon as the balance is redressed, and the market becomes bloated with an over-abundance, then quality becomes, to use the jargon of the Buying Officer, the primary purchasing determinant.

Why do manufacturing organizations adopt the principles and techniques of a Statistical Process Control and Total Quality Management? Is it because the boss thinks it will be a Good Thing to do it? No, only very rarely. The main reason is *customer pressure*. Do it, or else. Take it on board or lose our business. This is the quality imperative.

So it was with Bedlam plc. One of their major customers, NOVA Cosmetics, a world-leading toiletries and cosmetics manufacturer, had in desperation decreed that if Bedlam, as a principal supplier of decorated packaging, did not improve its abysmal quality performance, then all contracts would be cancelled, no more orders, no future with us! Do it right or die.

Which is how yours truly got onto the payroll. If Bedlam's plastics operation were to survive then NOVA's business had to be held on to. It was my job to make certain we held on to it.

'Your job is to make certain we hold on to that business,' barked the Big Boss. He was called 'Big Boss', being a glaring and burly fellow and MD of the whole company. The company was part of a bigger group.

'Hold *on to* it!' echoed the Little Boss. He was called 'Little Boss', being a narrow-eyed and wiry little chap and

Director of the plastics business. He dwelled in mortal
terror of Big Boss. He was sometimes referred to as Little
Sir Echo because of his sycophantic tendency to repeat the
Big Boss's proclamatory statements. This he practised as a
wrath-evasion tactic.

'NOVA are being awkward,' said Big Boss.

'Awkward, very,' Little Boss nodded in agreement.

'More than half of what we send *they* send back. For
quality. What kind of customer do you call *that*?'

Little Boss, unsure of the prescribed answer to this
rhetorical question, pondered with pursed lips and a frown
of concentration, to prove he was thinking about it. He
found the answer and said it like a question, 'Awkward?'

Words were already moving in circles. It was one of the
quickest entries into that system of circularized futility
known as a bottle-to-bottle hook up (this is explained at the
end of this chapter) that I had ever witnessed. Talk about
who's crazy!

'Exactly. Sending stuff back. Rejected by their Quality
Manager and he's *only* a quality controller, but he's making
our life a misery.'

'An absolute misery.'

'Now *you* are the new quality man, are you not?' Big Boss
leaned over the ramparts of that piece of furniture of
intimidation he called his desk as if challenging me to deny
my position. 'Your job is to make certain we hold on to
NOVA's business. *Stop 'em sending stuff back.* They're
driving us mad. Now what I want you to do is to write me a
report telling me *what* it is you intend to do about it; *how*
you intend to do it – full details; *when* you intend to do it –
full time-plan of sequence of actions; and I want your
proposed action plan on this desk within one week. Got
that?'

'One week. Got that?'

All very decisive stuff. Genuine Management by Objectives format as taught in the best business academies.

I ventured a question. 'Do NOVA know that I'm now on
your payroll?'

'Oh yes, of course they do, we told them we'd recruited
the right person for the job. But for them you wouldn't even
be here.'

'Be here.' The words rebounded off the barren escarp-
ments in Little Boss's skull.

'Do you think they expect me to achieve some sort of
quality improvement fairly soon?'

'Oh yes. Very soon. Like yesterday!'

'Like yesterday.' The echo could have irritated a sensi-
tive listener, but you soon learned how to blank it out.

'Then instead of writing a report telling you how I'm going to do it, why don't I just do it?' This was heresy in their corporate mythology. In asking it I was questioning their entire system of thought, their mindset, their culture. This is perceived as a very serious threat to the status quo.

'But we need to know what it is you're going to do and how you plan to do it, in writing,' commanded Big Boss. 'That's the way things get done round here.'

'Get done around here.' (Or don't get done?)

How I do it, what I *do*, is no concern of yours. None of your business, of no professional interest to you. You should *never* ask what does Aaron Godman do, ask *only* what does he *achieve*. Trust me.'

This, to them, was a novel approach. They did not want to *trust*. They wanted to *manage*.

'Trust you?' The lynx-like eyes of Little Boss held mine for a moment before flickering down to his doodle-pad again. You can read men's hearts through their eyes, once you know how. Here I was with these two bossmen, one blustering, the other wary, and both afraid. Afraid of what? That does not matter, fear can be projected on to anything, because it lurks in hearts. Oh, brothers, *why* do you feel it necessary to be afraid? There is no need of it.

'But what are you going to do?' demanded Big Boss.

'Yes, *what*?' asked the echo.

'I am going to do one thing,' I advised them, and kept talking to block out any echoing interruption, 'I am going to cultivate a climate within which people will strive for excellence, in order that excellence will flourish.'

'That's a *mothering* objective,' said Big Boss contemptuously, thereby consigning about half the world's adult population to a life of pointlessness. 'Anyway, where do you even start with a thing like that?'

'Where do I start? I have started right here.'

As I said before, the task facing your Gnostic 'priest – for ever after the Order of Melchizedek' Aaron Godman was not going to be easy. But it was going to be fun.

Concepts of Folly and Futility: The Bottle-to-Bottle Hook-up

There is an episode in Joseph Heller's anti-war novel *Catch-22* which eloquently captures and typifies the philosophy of Old Order management. It describes how the anti-hero of the narrative, Yosarian, after having been in a coma, regains consciousness in a hospital bed and becomes aware of the patient lying in the next bed. This patient is completely encased in plaster of

Paris, from top of scalp to soles of feet. Above the patient is suspended a bottle of liquid, connected to the head-end of the patient by a rubber tube. A second tube emerges from the patient's plaster carapace in the lower abdominal region and is inserted into the neck of a second bottle standing on the floor beneath the patient's bed. During the course of the day the liquid in the upper bottle gradually dribbles down the upper connecting tube until this bottle is empty. By this time the lower bottle has filled with the liquid which has trickled down the lower connecting tube. At this point the nurse attends to the patient's needs by disconnecting the bottles from the tubes, *switching the lower bottle for the upper bottle*, and reconnecting them to the tubes so that the input/output cycle can be repeated. Our anti-hero watches this daily ritual with mounting interest, then, being a lateral thinker and a minimalist, on the third day suggests to the nurse how the ritual might be refined. He recommends that the plastered patient is redundant to this cyclic system, and that all she needs to do is *connect the upper bottle directly to the lower bottle*, thereby bypassing the plaster of Paris sarcophagus which is really cluttering up the economic elegance of the flow-system. He thus creates a perfect bottle-to-bottle hook up.

A bottle-to-bottle hook up is a self sustaining cyclic system of ultimate futility; it is the delusion of perpetual motion, going nowhere; it is a ring-a-ring-of-roses in which nobody goes any-where and then everybody falls to the ground; it is activity masquerading as work; it is performed regularly up and down the realm of manufacturing, for instance, a case study:

'*Play it again, Sam*'. The factory makes things out of thermo-plastic material. The beauty of working with this kind of raw material is that if you make things wrong the first time round you can recover the raw material and have a second shot at it; if that also goes haywire you can do it over again, and again, again . . .

When asked what the quality costs in his plastics-extrusion operation happened to be, the company's Financial Director claimed them to be 'two per cent'.

'Two per cent of what?' the consultant enquired, 'two per cent of total sales value of production?'

'That's correct,' nodded the FD.

'Remarkable,' said the consultant. 'Truly remarkable for three reasons. Firstly, such a *low* figure in an industry whose typical costs of poor quality are anywhere between 15 and 40 per cent of sales revenues. Secondly because you seem to know the figure so *exactly*; how can you be so sure?'

The FD interjected, 'Because that's the figure that Birmingham Bill the scrap plastic merchant pays us for all our junked regrind plastic, hard cash, two per cent.'

'But thirdly,' went on the consultant, 'you are routinely and regularly regrinding *30 per cent* of all the plastic which goes through your extruders. So this surely represents a cost of poor quality figure much higher than your two per cent?'

'Of course it doesn't. You see,' the FD explained with the pitying exactitude which one might adopt when addressing an idiot child, 'what we do is take the rejected output, put it through the grinder to chop it into pellets, and then we put them back into the extruder. That way, you understand, we don't actually lose our raw material at all; except, that is, for the two per cent I've already told you about.'

The consultant, feeling himself being treated as if he were some kind of imbecile, and thinking he might as well carry on this dialogue of the daft, nodded, 'I see what you mean, and you have just pointed the way to the solution of all your manufacturing problems.'

'Have I really? That's good, tell me more.' The FD took the bait.

'Well, what you do,' the consultant advised with deadpan seriousness, 'is to build a conveyor from the extruder directly to the grinder. Then you build a second conveyor from the grinder to the extruder. All your output travels down the first conveyor into the grinder, then all the regrind travels on the second conveyor back into the feed-hopper of the extruder. Then – ergo! – you have created a perfect bottle-to-bottle hook up: no more scrap problems, massive reduction in raw material bills . . .'

'That's a crazy idea! Do that and we'd be out of business.'

'Is it?' the consultant persisted. 'It is actually a manufacturing system called Closed Loop Automated Process To Recycle All (or Any) Product. Claptrap for short. If it makes any sense to regrind 30 per cent of your output, as you are currently doing, then it makes *better* sense to regrind *all* of it. Or it makes no sense at all to regrind any of it.'

The FD accepted the logic of this proposition, and his company nowadays regrinds considerably less. And before we scoff at any company which has locked itself in a bottle-to-bottle hook-up of making wrong first time and then making over again, in the belief that this work which adds cost is somehow free, it might pay us to look around our own organizations. We might discover that we too are operating a few of these cyclic systems of futility.

Bedlam Plastics were circle dancing in bottle-to-bottle hook-ups. Are you also a circle-dancer?

Summary

Creating constancy of purpose, according to Deming's first point. The importance of vision becoming mission through sharing it by 'giving it away'. You can only ever give to yourself. The pre-eminence of leadership over management. The true nature of total commitment to total quality, a farmyard fable and two case studies from manufacturing industry.

The changing world of work. Responding to change, the four options – do nothing, do anything, or do the right thing. The proper use of knowledge and the slippery nature of language. What is, or is not, quality. The myths which obscure the true meaning of it. The goal of the organization – money – and quality is about helping to achieve that goal.

Bedlam, Aaron Godman's first task.

Concepts of folly and futility – the bottle-to-bottle hook up – making over again instead of right first time, doing work which adds cost instead of work which adds value.

3 DEMING'S SECOND POINT

> 2 Adopt the new philosophy. We are in a new
> economic age, created by Japan. Transformation
> of Western management style is necessary to halt
> the continued decline of industry.

This is Deming's Second Commandment. As with his first
Commandment, there is no arguing with it, but like any other
doctrinaire statement it calls for interpretation.

What is the 'new philosophy' he is urging us to adopt? Come
to that, what is the 'old philosophy' he is suggesting we should
discard? What is this 'new economic age'? What kind of 'transformation of Western management style' are we being advised
to consider?

Once again we find ourselves guilty of using our words upside
down. The 'new' philosophy is in truth about three thousand
years old (and there is documentary evidence testifying to this,
so it is very likely even older), and the 'old' philosophy is
perhaps less than three hundred. Herein lies the power of
changing from the 'old' to the 'new', and here is why . . .

The 'Old' Philosophy

This is the way we do things today. This is *Old Order* management, its assumptions and practices bequeathed to us by the
founding fathers of the industrial revolution and the originator
of scientific management – Frederick W. Taylor, the man who
slaughtered work and dissected its carcass into thin strips. In

the mindset of this Old Order people are perceived at best as essential, but not really wanted, extensions to the machinery of production; at worst as unpredictable nuisances to be tolerated until they can be automated out of existence. The Old Order's dark vision of the future is in the workless factory run by robots controlled by an elite of technocrats; in the meantime, until the full dawning of this manufacturing millennium, technocratic management will be obliged to continue managing its workers, as best it can, by the time-honoured (or time dis-honoured) carrot and stick system of reward and punishment. This is a managerial philosophy akin to the training of dogs, but we Old Order managers apply it to people. 'What other way is there?' we ask. In this view of the scheme of things people are believed to work for money (if they have no easier way of getting their hands on some cash that is), and it therefore assumes people will do as little work as they have to for as much money as they can get.

This opinion is correct. The managers who subscribe to this concept of the nature of man-at-work and woman-at-work know, from their daily experience of conflict and shopfloor militancy, that this view of the indolent avarice of the workforce is true. This is their reality.

Of course it's true. Under such a demeaning system of the managing of work, it is not possible for there to be any other truth. Treat people as if they were lazy dogs and they will respond accordingly. So, this pernicious misunderstanding finds continuous reinforcement in managerial perceptions of shopfloor behaviour. Because, as ever and as always, *you see what you believe.* Management, believing that collectively blue collar men and women are the embodiment of all that is bad in the workplace, because their experience to date tells them this is so, find daily confirmation that this is indeed the case. So conflict is made certain. People work for money, says the creed. The manager's job is to get as much work out of the worker for as little pay as he can get away with. The worker's ambition is diametrically opposed to this. This is the black and white world of Simple Simon thinking. This is the world of the Old Order management which we have today. This is the world we chose. We always choose our worlds. This one at least has the dubious merit of certitude, it harbours no doubts about the nature of work and why people do it. People work for money! So be it, say the taskmasters and the coolie-whippers of authoritarian management, unshakably confident in the rightness of their cause and the correctness of their actions.

Until the behavioural scientists and industrial psychologists hark upon the scene. They sing a different song, they tell us that people work for something other than money. They speak of 'satisfiers' and 'motivators', they confuse us, they open little cracks of doubt in that granite edifice of absolute certainty into which our experience chisels the eternal truth that people work only for money. They make it sound almost like a different sort of labour in a different world of work. When they use the word 'work', can they possibly be referring to that human activity which we have always called 'work'? Their definition of the word seems to be utterly at variance with ours . . .

It is. Because there are two kinds of work!

We have already seen that there is work which adds value and there is work which adds cost. This is a legitimate division of work into two distinct categories as seen from the standpoint of factory accounting.

There is a further dichotomy, seen this time from the viewpoint of managerial philosophy.

There is work which meets a person's deficiency needs (D-work), and there is work which meets a person's Being needs (B-work).

D-work is about earning a living. B-work is about enjoying life.

D-work is resented, but basic. B-work is sought after, and exalting.

Deficiency is about nothing more than making good a short-fall, filling a gap. People need money to survive, so they work for it. People who lack vitamin C develop scurvy, so they eat oranges. Once people have earned a few pounds or dollars they are able to survive, which is not the same as living. When the scurvy patient has sucked a few oranges his symptoms subside; he is no longer dying of scurvy but by no stretch of the imagination could you say he was enjoying robust health and a fullness of being alive. D-needs do no more than drag you back from the edge of the grave; to really live you have to fulfil your B- (Being) needs. This knowledge presents us with a golden managerial opportunity. We shall explore it in depth later on.

The two kinds of work are as different as the two sides of a coin. Old Order authoritarian management looks upon work only from the D-standpoint. The New Order observes it from the B-vantage point. That is the only difference. It is a small difference, yet each is a world away from the other. Since all our actions spring from our beliefs the behaviours engendered by the differing standpoints are also worlds apart.

We are indebted to Abraham Maslow for having invited us to see the world of human need and work from these two observation posts. Let us look further into his vision.

WHISTLE WHILE YOU WORK

How do you *feel* about your job (i.e. about your work)? If the company assured you that whether you attended your place of labour or whether you chose not to do so, *they would pay your wages into the bank regardless*, would you actually bother to go into work? Or would you give it a miss and carry on drawing company money for nothing? (Well, for all anybody knows you might be doing that already, by being at your place of work in the flesh whilst being absent in the spirit. Some people have spent the better part of their lives in this state of escapist dissociation.) What would you do? Would you go into work?

If your answer is 'No', the chances are you are working mainly in the D-realms (Old Order); if 'Yes', you are probably spending most of your time in the B-realm (New Order).

Now please try another one. All of us experience emotional 'ups' and 'downs', get turned on and elated, or turned off and depressed, by the things that happen to us at work. Cast your mind back to some occasion or situation at work when you felt really good, really exhilarated and switched on. What induced it? A promotion? A pat on the head for a job done well? The successful completion of a challenging task? The company's adoption of some recommendation you made that caused things to go better than they had been going? A sudden and unexpected pay increase? The joy of helping one of your subordinates get elevated to the same rank as yourself? Successfully sticking a spanner in the works of a political opponent? Or what? If you now care to look at Maslow's ladder (see Figure 3.1)

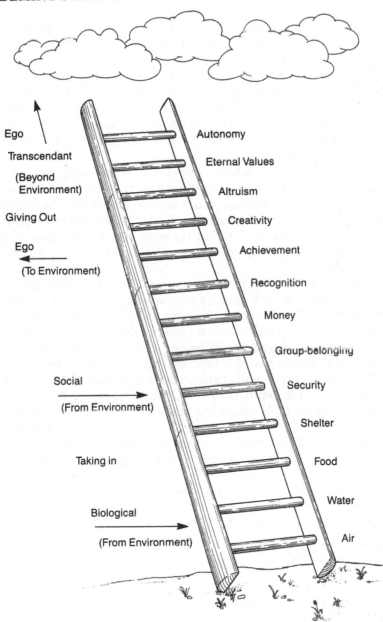

FIGURE 3.1 THE MASLOVIAN LADDER OF HUMAN NEED

you will be able to assign the work situation which prompted the positive feeling of elation to its appropriate rung. This is the rung you are living on, in your work, and probably in your non-working time as well, because this is the kind of person you are choosing to be in the context in which you are currently operating. Are you on one of the 'social' rungs? Or the 'ego' rungs? Or the 'transcendant'? If your position on the ladder is lower than the ego rungs you are most probably working at D-levels. What is holding you there?

What about your subordinates? On which rung are you shackling them? Or are you helping them to climb this ladder of instinctual need? And what about the workforce, the blue collar men and women who generate all the added value in your organization – are you holding them on the 'money' rung and then complaining about their greedy demands for more pay for doing less work?

If you are holding them down, you are holding yourself down.

You can only grow bigger by encouraging the growth of your brothers and sisters at work. You cannot climb the higher rungs of the Maslovian ladder on your own, to do this you have to be prepared to take them up the ladder with you. The higher they climb, the higher you do. (We are not speaking here of the artificialities of rank and status, we are speaking of access to a fuller degree of humanness.) All this runs counter to the conventional wisdom of Old Order management. In that land of dog-eat-dog one person's gain is seen to be balanced by another person's loss. Either you are one up, or you are one down. Either you win the game or you lose it. This is a black and white world of absolutes, there is no colour in it. It is a place unfit for human habitation. Yet it is a place where many of us seem to be required to pass our days in pointlessness. It is a miserable hole, a madhouse. The New Order offers us a better world than this.

The 'New' Philosophy

This is the way we shall be doing things tomorrow. Some of our more enlightened organizations are consciously seeking to begin doing things this way today. This New Order of management is far older than the so-called 'old philosophy' which Dr Deming speaks of; it is as old as (probably older than) recorded history.

Unlike the Old Order assumptions which are job-description based, and are concerned with trying to bend the person to fit the job, and therefore to constrain and control him or her, the New Order approach is more concerned with flexing the job to fit the person, to liberate and lead. So where the Old Order is mechanistic, the New is consciously humanistic, and in being so is able to draw its inspiration from deeper wells of insight and understanding than that afforded by the poisoned shallows of Old Order thinking. This is the fundamental reason why New Order leadership is proving to be so overwhelmingly effective wherever it is beginning to be practised. It is a reaffirmation of humanness in a world of machines. Above all, *it works*.

New Order thinking is the foundation of the 'new economic age' referred to by Dr Deming. It is about getting more effective performance out of a workforce by controlling them less. This is a paradox. New Order thinking is rich in paradox – what could be more paradoxical than 'giving in order go get'? What could be more paradoxical than 'obtaining a firmer grip by letting go'? Yet this is part of the philosophy of manufacturing which has led Japan to its position of pre-eminence, and brought about the new economic age of global competition.

The 'transformation of management style' which Dr Deming urges us to adopt means we must learn the precepts of New Order management and thereafter act upon them. There is nothing difficult about transforming our management style: we are free agents, we have free will, we can adopt whatever style we wish. We can choose.

Or can we?

The Idiot's Tale: 3
When you are deep in the woods, fighting bush fires, it is not surprising that you are unable to see the shape of the forest and how it relates to the topography of the surrounding landscape. Bedlam were in this position, especially in their relationship with NOVA Cosmetics. Firefighting was their way of life. Whenever fires are roaring firemen are highly regarded, and even though the flames go higher the firemen are applauded. Dashing about with a sweat-streaked face amid smoke and sparks is an eminently visible *activity*, and activity generally finds its reward. The hyperactive are promoted. This is called management succession planning.

Bedlam were actively managed. Bedlam were burning. Bedlam's managerial atmosphere was a choking mixture of the smoke of deceit, which served as a screen to obscure management incompetence, and the hot air exhaled by managers shouting at supervisors, supervisors shouting at the workers, and the Little Boss shouting at everybody. Old Order organizations can be noisy environments in which stridency of voice becomes an indication of rank; so Little Boss always shouted loudest except when Big Boss was around, and he shouted loudest of all. All very British.

'There's nowt but a load of idiots working for me,' was how little Boss explained the company's abysmal performance; 'can't do owt right.' (He pronounced it 'rate' in the patois of his people, in order that his message should be clearly understood.) The managers had other ways of explaining poor performance. 'The machines are clapped out,' they diagnosed. 'The raw material's no good,' they averred. 'The workers are not committed,' they criticized. 'We've got too many problems,' they complained.

'You have only one problem,' I suggested to them, 'and it is in your heads.' This kind of comment is not of the Dale Carnegie How to Win Friends and Influence People sort, but now and again, observations like this have to be made by somebody, so why not by me, the newly arrived enigma, the new Quality Man? And Bedlam blazed.

What is the ancient antithesis of fire? Water. Imagine that *knowledge* is a fluid as precious to the mind as water is to the body: and imagine further that a vast reservoir of this liquid is out there somewhere waiting to be tapped.

How do you bring water to fire? You construct canals, conduits, aqueducts, ditches . . . a *network* through which the quenching life-bringing liquid can be conveyed. How do you convey knowledge? In the same way, through a network of *communication*.

Communication: now *there* is a word which has been as brutally misused as any. 'We are communicating,' yelled Bedlam's managers, when in fact they were doing the opposite. Like the generality of Western managements they talked a lot about the need to communicate but never actually talked with anybody; they harangued, pleaded, bullied, beseeched, bemoaned . . . but they rarely said anything. Why? Because they had nothing much to say.

It was time to tap the reservoir. Time to smite the rock with a staff and cause the shining water to bubble forth. To do again the age-old miracle. Did you think the age of miracles had died? Perhaps the obvious will one day dawn upon you, when you discover that you have been living in an age of miracles all the time, but it somehow escaped your notice, and you have remained the poorer for it.

The time to bring water to Bedlam's blaze had arrived. But to build a communications network of canals and conduits with which to draw upon the limitlessness of the reservoir takes a little time. Bedlam were running out of time as swiftly as their prime customer, NOVA Cosmetics, was running out of patience. A faster way of bringing the water to the fire had to be found. What moves water faster than a canal? *Rain* does. Rain, like mercy, drops from above onto everyone beneath. So somebody had to make it rain. Who? The Rainmaker. Who is the rainmaker? Why, I, Aaron Godman the enigma. Who else? Anyway, rain-making was one of the subjects covered by the divinity curriculum of my lesser-gods apprenticeships scheme. Rain making is a miracle, and like all miracles, it is easy . . .

Colin Wilson recounts in his book *C. J. Jung, Lord of the Underworld* the story of the Rainmaker, as it was told to Jung by Jung's friend Richard Wilhelm.

> Wilhelm was in a remote Chinese village that was suffering from drought. A rainmaker was sent for from a distant village. He asked for a cottage on the outskirts of the village, and vanished into it for three days. Then there was a tremendous downpour, followed by snow – an unheard-of occurrence at that time of year.
>
> Wilhelm asked the old man how he had done it; the old man replied that he hadn't. 'You see,' said the old man, 'I come from a region where everything is in order. It rains when it should rain and is fine when that is needed. The people themselves are in order. But the people in this village are all out of Tao and out of themselves. I was at once infected when I arrived, so I asked for a cottage on the edge of the village, so I could be alone. When I was once more in Tao, it rained.'
>
> By being 'in Tao and in themselves', the old man meant what Jung had meant by individuation. That is to say, there was a proper traffic between the two selves – or the two halves of the brain. The people in the rainless village were dominated by the left-brain ego – which, while it is unaware of the 'hidden ally', is inclined to over-react to problems. This in turn produces a negative state of mind that can influence the external world.
>
> This throws a wholly new light on the idea of synchronicity, and also of magic. One could say that, according to the Chinese theory, the mind is intimately involved with nature. Synchronicity is not therefore the active intervention of the mind in natural processes: rather, a natural product of their harmony. (So when we are psychologically healthy, synchronicities should occur all the time.) Our fears and tensions interfere with this natural harmony; when this happens, things go wrong.

Things go wrong! Things were going wrong in Bedlam Plastics; there were problems, and like the people in the

rainless Chinese village Bedlam's managers were over-
reacting. Locked in the ego mode of left-brain thinking
they endlessly 'analysed' their problems. The more they
analysed them the bigger they became. So they analysed
some more, and the problems seemed worse than ever. As
head of the operation the Little Boss lamented the vile
misfortune that had burdened him with so many fools for
subordinates, so he shouted his disapproval and angry
disappointment into their ears. They in their turn bullied
the supervisors, who swore at the workforce, who sullenly
cursed their fate, their bosses and life as it is lived. These
people were out of Tao, out of themselves, out of order.

Order had to be restored. But it is not easy to bring about
order to those who are not only accustomed to disorder,
but look upon it as 'natural order', and embrace it, like a
reassuringly familiar spouse, so as to be able to continue
complaining about it. Disorder has its attractions – it is the
legitimized drunkenness of that socially approved figure,
the 'workaholic'. Big Boss was a workaholic, and proud of
it; he put in long hours. Therefore Little Boss, in faithful
reflection of the role-model, was another workaholic; he
put in longer hours. So their subordinates, adopting the
protective coloration of self-preservation, were also workaholics, and they put in the longest hours of all. A workaholic is a man occupying a position of minor authority in
some kind of hierarchy who is *hooked on deference*. Like
anyone addicted to any kind of drug he cannot get enough
of it, the more he gets of it the more he craves for yet more.
Deference – the obsequious and fawning response to authority – it is a dangerous drug. There should be a Government health warning issued with the allocation of authority. The workaholic finds the drug of deference at work,
where he is empowered to lord it over subordinates; he
finds precious little of it at home, none from his wife and
just a little from the kids until they grow old enough to
know better and cut off his supplies. So he spends more
and more of his time at work. Not because his job requires
it, but because his addiction demands it. Workaholism is
an industrial disease. It numbs the brain. It is a conspiracy
of deceit foisted by men upon wives. It inflates the ego. It is
the wrong reason for taking a job. It drives out compassion.
It kills creativity. It tramples on justice. It is very highly
regarded.

Nevertheless, workaholics or not, order had to be
brought to Bedlam. Order had to be brought also to the
tetchy relationship between Bedlam and their major customer, NOVA Cosmetics. Where discord now ruled con-
cordance and harmony had to prevail. If there is not

enough time to build the formal channels of communication so dear to the heart of the bureaucratic manager, then
communication has to be achieved in a quicker and
simpler way. How? By talking with people, actually talking,
having something to say, and saying it.

You see, it's all very well constructing the formal channels of communication, digging the canals that lead to the
noticeboards, contriving the conduits of the 'briefing
groups', raising the aqueducts of video transmission to
project the message of the Big Boss on the canteen wall . . .
these channels are only the *structure* of the communication system, it's what you pour into them that counts, the
content. If you have very little of any value to talk about
then the canals will run as dry as those on the planet Mars;
only rusty droplets will trickle out of the conduits, to
disappear into the sand.

So I talked to them. All of them, at Bedlam and at NOVA.
Talked about quality. Told them what it is really all about.
Relaxed them, quietened things down, showed them that
work can sometimes be fun. Made it rain. Quality performance began to improve, not in a matter of weeks, in a
matter of days. Big Boss didn't feel at all comfortable with
what was happening.

'What's happening?' he asked, nudging his blotting-pad,
whose position had been disturbed by the office cleaning
lady, into its correct alignment on the universe of his
desktop, where every article occupied its ordained position
of latitude and longitude as exactly as the heavenly bodies
are set for ever in their proper places in the celestial sphere.
This was his realm of ultimate order, this desktop, whose
very pens stood to attention in their wells, whose onyx
cigarette box sat squat as a concrete gun emplacement
guarding its particular and precise corner. All was order on
this acre of mahogany and green leather, an island of order
in an ocean of chaos. The Big Boss adored his desk. He felt
safe behind it. In control!

'Quality's getting better,' I told him.

'Why?' he asked.

'Because it's time it did, I suppose.'

'What are you *doing* about it?' he persevered down the
wrong track, so I thought I might as well be a bit more
'enigmatic' and taunt him a little, he seemed to expect it of
me by now.

'Practising the art of masterly inactivity.'

Masterly inactivity,' he exploded, 'sounds more like
bloody idleness to me.'

'NOVA Cosmetics seem to like it,' I reminded him,
cruelly perhaps, because at the mention of the dreaded

customer's name his eyes narrowed a little, like a boxer's
when he's evading a blow. I pressed home the advantage. 'If
what we're doing, whatever that might be, suits the customer and our company accountant, then it should suit
everybody, I guess.'

'Mm,' he pondered, 'that's all very well, but I want to
know what is happening, *what's going on*?'

'Boss, don't ask how the water gets out of the rock when
it is hit with the staff, just drink and be grateful.' More of the
enigmatic stuff.

'There should really have been no need for the Big Boss
to ask what was going on. It was happening all around him,
in his full view. Trouble is, though, that quality is an
invisible input, done by invisible people, employing invisible means. Invisible to those who have no eyes to see,
that is.'

So, what *was* going on?

A reversal in quality philosophy, from defect-detection
to defect-prevention, was going on. This switch in managerial thinking is a painful transition for some people to
make, but unless it is made nothing else changes, and
quality stays in crisis. This is how it was brought about in
Bedlam . . .

1. WHAT DOES THE CUSTOMER REALLY WANT

Explicit and unambiguous specifications were negotiated.
Defects were defined and classified according to degree of
severity. This was done in the light of British Standard
6001, which is based on Defence Standard 131A, derived
from American Military Standard 104D.

Acceptable Quality levels for each class of defective were
agreed. It was acknowledged that whilst nobody *wished* to
make defective output, it might nonetheless occasionally
arise, if for no other reasons than the uncertainty principle,
the laws of chaos, or simply Murphy's Law, so the concept
of zero defects was regarded more as a target – a guiding
star – than as an operational destination.

Systematized sampling of the product, with agreed
batch-sizes, was instituted at the plant.

These standards were then rigorously and ruthlessly
enforced.

2. CAN WE MAKE IT OK?

This was the pain barrier of the switch in quality philosophy. Defect-detection quickly did its work, and amassed a
mega-heap of rejected batches into quarantine by prevent-

ing shipment of non-conforming product. This was a pain-ful experience for Production Management.

Now came defect-prevention. *Why*? became the prevail-ing question of the day. Why are we making so many defectives? Why are we making any defectives at all? What is the capability of the process and how does it relate to what the customer wants?

The asking of why? led on to the asking of how? How can we improve this or that process so that it produces pristine quality at all times? This triggered-off controlled enquiry into the capabilities and the vulnerabilities of the produc-tion system and its people.

So a process of *education* was set in train. People began to work smarter instead of harder. Realism replaced wish-ful thinking. The status quo was challenged.

Attitudes were altered. Morale was raised.

Communication was the key. The customer, hitherto seen as a demanding nuisance, began to be perceived as a partner in a supply-chain of co-makership. A continuous dialogue of quality was maintained, you could say that the quality improvement was 'talked-in'.

'All he ever seems to do is talk,' acidly commented Big Boss. But the talking did its work.

Quality performance continued to get better. Through the application of well-tried and well-proven statistical techniques, as recommended by Dr Deming and available from a host of sources, Bedlam Plastics followed the precepts of Right First Time manufacturing. In less than a year, when NOVA Cosmetics did their annual Vendor Assessment, Bedlam had moved from supplier position number 36 (NOVA had only 36 suppliers, otherwise Bedlam would have had a worse rating) into supplier position number one.

As Dr Deming says in his second point, 'transformation of Western management style is necessary to halt the continued decline of industry', and that kind of transfor-mation began to happen in Bedlam Plastics. Decline was averted. Customer dissatisfaction was turned into mutual trust. Operational losses due to bad quality practice were reversed into measurable profits, all amply proving the validity and relevance of Deming's second point. Old dis-Order management metamorphosed, like a dry chrysalis into a winged creature of aerial beauty, into New Order. Because the climate changed. Because it rained.

Summary

Unless Western managements change their style, moving from the traditional Taylorized way of ordering things, to a more enlightened view of the nature of people and work, decline will continue. The mechanistic approach to managing people is unable to liberate and harness talents in the workforce which are being so successfully mobilized in Japanese organizations. This 'new philosophy' is the trigger to activate all the people on the payroll to a higher level of personal commitment and motivation.

This new enlightenment demands more of employees than the old order ever did or could. It is humanistic. It is a different kind of work, based on Maslovian and other concepts of human need.

This approach was adopted by Aaron Godman in Bedlam plc, to transform a crisis into a success. This new way of ordering work seems, to some people, management by mysticism. It may well be, but *it works*. Its success is measurable in the terms which everyone can understand – measured improvements to profits performance. The key to its success is *communication*.

4 DEMING'S THIRD POINT

> 3 Cease dependence on inspection to achieve quality. Eliminate the need for inspection on a mass basis by building quality into the product in the first place.

Try this: Select any page of this book (or any book) at random, count the number of letters 'r' – both upper and lower case – appearing on the page. Note how many you have counted. Now pass the book to a colleague and ask her to do as you have just done – count the 'r's – and to make a note of how many she finds. Once more, hand the book to another colleague and invite him to repeat the exercise. Now each of you has written down a number which states how many 'r's there are on that page, compare your results; they are all different! How many 'r's are there really on that particular page? Who knows for sure?

You have just conducted a mass inspection on a batch of work (a page) containing about 1,500 to 2,000 items (letters of the alphabet), searching for rejectable units of output (the letter 'r'), and when you compare the results of three inspectors (your two colleagues and yourself) you discover that none of you knows how many rejects there are in the batch. So it is impossible to state with any degree of accuracy the percentage defective of that batch. Furthermore, if you had been screening out letters 'r', pulling them from the text, some would have eluded your inspection and would have remained in the batch.

This is how mass inspection works: or rather does not work. Which leads us to a Golden Rule:

Mass inspection always lets you down.

Even if you had actually erased from the page all the 'r's you found, your colleagues would have found some of those that you had missed when she did her inspection of that page, she would have erased these but those which she missed would have been there to be discovered by the third colleague, and he would not have found all of those either, so even after three inspectors (300 per cent inspection) have done their best to screen duds out of a faulty batch, it still contains a residue of undetected duds.

This is bad news to any manufacturers who believe in the Old Order 'make it – inspect it – send it' style of management. If 100 per cent inspection is not effective, goes the rationale, it must be because the inspectors are 'lacking vigilance' or are 'not committed' or are 'lazy and inattentive' . . . so the inspectors get the blame for bad work remaining in the output. They are actually being blamed for being merely human, and therefore fallible.

Yet this is how 'quality' is arranged in those of our organizations still operating under Old Order thinking. Managements fail in their duty to build quality into the product, then discover they have made large batches of output contaminated with too high a percentage of defectives, so they have these batches mass-inspected by teams of workers usually called 'inspectors' but in reality performing a filtering out, or 'sorting', process, and then complain about the workers' capabilities when it is discovered that too high a number of defectives still remain in the 'sorted', batches. *Quality cannot be inspected into the mass of output by screening duds out*; to attempt to do so is like contaminating a river and then trying to filter out the contaminant; better to have prevented the contamination in the first place. (This is discussed later in this chapter under the heading Concepts of Folly and Futility – The Poisoned River.)

So if we cannot inspect quality into the output by screening rejects out of it, we are *obliged* to follow Deming's third point and build quality into the product in the first place, an ounce of prevention being worth a ton of detection. But how? By finding the *cause*, and eliminating it.

Ah, there's the rub. Easier said than done. This is where knowing the technology, 'sticking to the knitting', comes in. Each quality 'problem' which is in fact a symptom of some underlying disorder brings with it its own quality 'solution'. These problems and their solutions are as different as the technologies within which they arise. That's the bad news. Now for the good. The statistical tools of quality are universally applicable, no matter what the technology happens to be. Just as you can use carpenter's tools to shape a few slats of soft deal into a bird-table, or to fashion seasoned planks of bone-hard elm into a boat's hull, statistical tools may be used to *transform the data* of any process from any technology into the *information* needed to ensure the achievement of built-in quality. Whether the problem is one of *attributes* (remember, an attribute is a yes or no thing, either it is present or it is absent, and attributes are counted rather than measured), or one of the *variables* (a variable being any characteristic of a product to which it is possible to assign a numeric value, for example, the weights of a succession of items from a production line, and variables are measured not counted), the statistical approach is the only one which makes sense and so the only one offering any hope of success. To illustrate this let us take a case study from the world of packaging manufacture, the production of tinplate cans:

You Know What Canz Meanz?

To paraphrase a well-known advertising jingle,

> A million housewives every day
> Open a Can . . . and throw it away.

After having emptied the contents into the saucepan of course. But throwing the can away without a second thought is the very best tribute it is possible to pay to the *quality* built into the can right the way back up the chain of supply. Everybody involved in the chain has been doing his or her job correctly, building in the invisible input – superlative quality, so the culmination of all their collective effort is that its outcome is taken for granted. This is exactly as it should be. If ever you notice the quality of a metal can, it's because it is of poor quality, it's blown. When did you last see a blown can? There are not many about, maybe a few: counted in parts per ten million, nearly zero defects, perhaps actually so, though we shall never know for sure. That's one of the uncomfortable things with a zero defect philosophy, you can never know when you have attained it, you tend to know only when you have not. Still, it represents an eminently worthwhile ambition.

In any multistage production process poor quality from earlier stages is detrimental to the quality achievable at a later stage. Furthermore, since the raw materials of one conversion process are the finished goods of a previous conversion process – that of the raw material supplier – the same condition prevails. Thus the imposition of high quality requirements at the conclusion of a manufacturer's operations demands the imposition of high quality standards on all preceding stages of production. This Right First Time approach, if it is to be successful, has to be enforced right back to the beginning of the production operation, and even further back into the processes of the raw material suppliers.

Can manufacturing, like any other high-volume production system which converts raw materials into finished goods, is unable to achieve 100 per cent perfection. This process, whose raw materials are less than perfect, whose hardware can never be flawless, is no more able realistically to aspire to the seductive fallacy of zero-defect output than is any other field of mass production. The concept of zero defects, like the concept of absolute zero degrees Kelvin, is beyond reach. But only just.

So the unrelenting torrent of output that is modern can manufacturing is inevitably tainted from time to time by the random occurrence of defective cans. If their incidence exceeds the agreed statistical quality levels, their presence jeopardizes the acceptability of entire batches of otherwise acceptable cans, and might result in massive customer's returns. At best, they will engender punitive in-house rejections leading to prohibitively high wastage. They are an expensive pollutant in the manufacturing stream.

We now need to look at the technology of can manufacturing, in order to appreciate the powerful way in which statistical methodology tackles quality problems (and therefore removes quality symptoms).

The Manufacture of Tinplate Cans

THE OLD METHOD

Tinplate canisters (abbreviated to 'tin cans' or simply 'cans') have been made since the early nineteenth century. Their method of construction required the rolling of a rectangular sheet of tinplate into a cylinder, forming the two opposite edges of the sheet into a longitudinal side seam by soldering them together so as to contrive a cylindrical body open at both ends. Each open end was then flanged in order to provide anchorage for the disc which formed the seamed-on bottom of the can, and the other end left open (giving rise to the term 'open-top' can) to enable the cannery to fill the can more easily and then to seam on the lid prior to steaming the can to cause sterilization of its contents. Consisting as it does of three components – the body and two ends – this type of can is known as a 'three-piece' can.

This method of manufacture endured for more than 150 years, undergoing steady refinement as materials and production methods improved, until it reached the ultimate of its technological evolutionary potential. Cans made by this method are produced today in their millions, more than 6,000 million a year in the UK alone.

One of the defects that such a construction is prone to is the splitting of the tinplate during the rigours of the flangeing operation, when the open end of the cylindrical body is rolled outwards and the increase in circumference arising from the out-turning flange imposes too great a hoop strain on the metal, which yields into fracture. Such a fracture would permit the ingress of bacteria after the canning operation, with consequent spoilage of the contents instead of their safe preservation, producing a 'blown' can.

The entire output of cans is accordingly tested by automated devices, and splits are rejected, to become line scrap. But even low levels of scrap, at high volumes of production, represent considerable financial loss.

THE NEW METHOD

It is in the nature of technology that it is constantly advancing – sometimes as a gradual operation, and occasionally in the form of a leap forward made possible by improved raw material behaviour or machinery capability. This kind of technological bound happened about twenty years ago in can making with the advent of highly ductile steels, and hence highly ductile tinplate (tinplate being steel lightly coated with metallic tin). Tinplate made from such new and accommodating stock lent itself to an innovatory method of making cans. This consists of deep-drawing the can from a disc of material, forming it under immense pressure into a seamless cylindrical container with an integral end (or 'bottom'). The walls of the can are then squeezed through circular dies so as to elongate, or 'wall-iron', them in much the same way that baker's pastry thins out under the pressure of a rolling pin, to produce a 'two-piece' container having the same volume as its three-piece equivalent but made from considerably less tinplate. Since the tinplate costs represents more than 60 per cent of the total cost of the finished can the economy of this system or production is self-evident.

Equally self-evident is the fact that this is a method of manufacture which imposes heavy demands on the properties of the tinplate, especially its capability of withstanding the thinning of the ironing operation and the rolling out of a flange – without splitting. It is asking a lot of the basic steel. Sometimes it is asking too much.

The Problem

As was said at the outset, it is one thing knowingly and willingly to produce defective output in the early stages of a production system on the grounds that defectiveness will be screened out downstream. It is another thing knowingly and *un*willingly to do so, and screen it out later.

It is another thing entirely knowingly and unwillingly to produce defectives upstream knowing that you are *un*able to screen them out in downstream operations. This is a potentially deadly trap. It is exactly the situation in which a leading can maker found itself during its early days of pioneering its draw and wall-iron process. It happened like this.

Tinplate is delivered to the can factory in ten-tonne coils, each coil about 30 inches wide and 0.012 inches thick. It is uncoiled and led through a press which stamps six discs from across the width of the coil at every stroke, instantaneously drawing them into shallow cylindrical cups. The cups are conveyed to machines which punch them through dies which reduce their diameter and increase their length so as to form them into the bodies of cans. At a later stage the bodies are flanged at their open end, at which stage they either split, or do not split. This company discovered that sometimes they did and at other times they did not.

If such split flanges occur at an incidence greater than 25 per 100,000 in any batch of cans, the entire batch is unacceptable to the customers. Inspecting to an Acceptable Quality Level as low as this by manual sampling methods demands a colossal deployment of costly inspection resources, but in the absence of reliable *automated* 100 per cent checking there is no other way of guaranteeing outgoing quality. Neither is there any way of sorting the bad from the good in batches rejected by the sampling scheme: result – high scrap losses.

Because of the apparent randomness of the arrival of split flanges it was necessary to batch sample-inspect the total output – time-consuming, wasteful and demoralizing. An automated testing device was on order, with a lead time for delivery of more than six months. In the meantime there was no way of screening out split defectives and no way of knowing when or at what level of incidence they might occur to pollute the output stream of an otherwise acceptable product.

A level of uncertainty such as this is intolerable in any manufacturing process. Something had to be done. Upstream contamination had to be prevented; a solution had to be found. This is how the can makers found it.

The Solution

The three rules of Systematic Quality Control are:

1   No inspection or measurement without recording;
2   No recording without analysis; and
3   No analysis without action.

But where to begin? What was to be inspected or measured other than the periodic and devastating arrival of splits in the flanges of some of the cans? This was already being amply recorded in full and melancholy measure.

Some other behavioural characteristic of the tinplate had to be assessed and recorded, and the results obtained from this set against the recorded arrivals of splits to establish whether a correlation (implying the possibility of a cause–effect relationship between the two factors) existed. But bear in mind – keep it simple.

Accordingly a simple test method was devised to enable quality control to measure the split resistance of the tinplate.

METHOD OF TESTING SPLIT RESISTANCE

One 'lift' of six drawn cups, taken from across the width of the tinplate coils, was converted into six can body shells on one selected bodymaking machine (to eliminate between-machine differences). These were then inverted, open end downwards, on to a steel cone of 60° taper whose base diameter was bigger than the diameter of the body shell, and compressed onto it. This compressive force flared out the open end of the body shell to the point of fracture of the rim, and the applied force at which fracture occurred was recorded (see Figure 4.1).

The output made from this coil was monitored and the frequency of occurrence of split flanges noted. This procedure was carried out on a succession of many coils. On the assumption that a thicker-walled can body would probably be more resistant to splitting than a thinner-walled specimen, wall thicknesses at their thinnest zone were also recorded.

On the assumption that the location of the split on the rim of the body might be of some significance, this was also noted, using the rolling direction of the coil (as indicated by the 'grain' of the metal visible in the can body) as a north–south reference line and allocating a compass bearing to the site of the split.

Whenever process information is gathered in a systematic manner it rapidly builds up into a picture that means something. Patterns of interaction begin to appear. Firstly there emerged the obvious and expected result, namely, that a coil which showed high values on the conic fracture test showed a correspondingly low incidence of split flanges during production (see Figure 4.2).

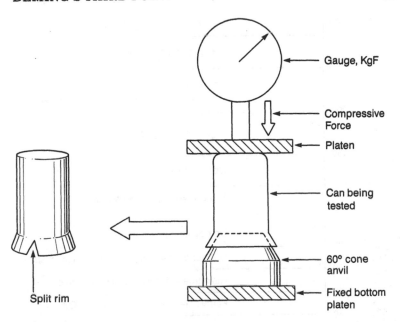

Gauge, KgF

Compressive Force

Platen

Can being tested

60° cone anvil

Fixed bottom platen

Split rim

FIGURE 4.1   CONIC FRACTURE TEST

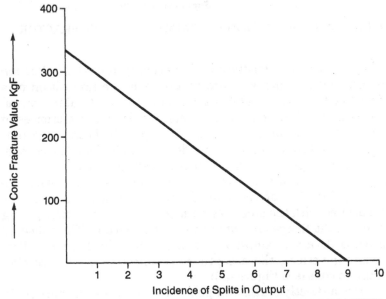

FIGURE 4.2   CONIC VALUE vs OBSERVED SPLITS (IDEALISED)

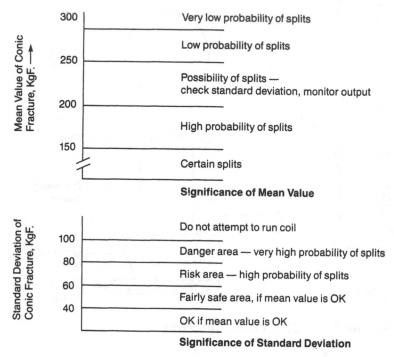

FIGURE 4.3   OPERATIONAL STANDARD – CONIC FRACTURE

The first and immediate action arising from this collection and analysis of process information was the imposition of a conic test fracture value as an acceptance standard on all incoming tinplate coils. The effects of thickness differences of the body walls of tested cans had quickly been ruled out; significant differences in the incidence of splits found no correlation with insignificant differences in wall thickness.

The location of splits was found to be associated with the rolling direction of the coil. There was clearly some factor associated with the steel itself which was intermittently giving rise to split flanges in some cans in the output. The collected process data was summarized into a Standard Quality Procedure governing the acceptance or rejection of all incoming tinplate coils (see Figure 4.3).

From a strictly statistical point of view, this procedure could be said to be incorrect because it embodies a deliberate conceptual error – to be explained later in the text. This procedure decreed that no coil should be run into the production line until it had been approved by the conic fracture test. The results were dramatic. The problem of intermittent split flanges in the output, with consequent high internal rejection of polluted batches, was all but eliminated. The stock of coils rejected and quarantined on account of their failure to meet the conic fracture standard piled up rapidly, to the dismay of the tinplate suppliers, who at first insisted that since the coils conformed in all respects to the prevailing selling specification they were acceptable to the terms of the business contract. It was emphasized to them that quality is fitness for purpose rather than conformance to an inadequate specification and that these offending coils were clearly unfit for their intended purpose, regardless of whether they conformed to a specification or not.

Raw material suppliers, like the suppliers of any goods or services, are naturally anxious that the things they sell should satisfy their customers, and remain sold rather than be returned as rejects. In pursuit of this aim, they are always willing to deploy expensive resources of research and development. So the facilities of the laboratories of the tinplate mills were enlisted in the investigation of the origin of split flanges. The in-plant problem of splits in the can factory had for all practical purposes been solved, but mounting coils on to a press only to have to dismount them after they have failed an acceptance test is not a profitable way of passing manufacturing time. Prevention at source is always better than a cure.

To appreciate how this preventive action was brought about it is necessary to make another digression – into the field of steel making.

STEEL MAKING AND ITS INFLUENCE ON CAN MAKING

The molten contents of a crucible of steel may be 'harvested' by one of two methods – either by being cast into large moulds and being left there to cool into ingots, or by being poured as a continuous stream into a tun dish from which the steel emerges as an extruding and solidifying ribbon which is then sheared into slabs. This latter is the more modern process.

In the former, ingot cast, process, a charge of aluminium is added to the molten steel. This induces suspended impurities to rise to the surface of the ingot where they form a crust of slag as the ingot cools. This crust, whose core descends a short distance into the ingot as a 'neck', is cropped off before the ingot is rolled into the coil which will later be electroplated to become tinplate. Any residue of the slag which is missed by the cropping operation remains in the ingot to be rolled into the coil into which the ingot is converted. Here it resides as a string of non-metallic particles, scattered within the body of the steel like currants in a cake. It is these non-metallic inclusions, which by their nature are unable to behave in the manner in which clean steel behaves, which give rise to fractures in subsequent deep-drawing processes.

In the latter, continuous cast, process, the puddle of liquid steel at the pouring tun dish is protected from oxidization by a thin covering of inorganic and inert casting powder which is steadily scattered during the period of the cast. If any of this is inadvertently drawn into the vortex of metal moving down the caster it remains within the body of the cooled steel in the same way as the inadequately cropped neck of slag remains in the ingot, with the same consequences – a high count of non-metallic inclusions resulting in a high incidence of splits during the can making operations.

It is during the conic fracture test of the coil of tinplate at the can-making factory that the presence of these non-metallic inclusions is detected. They might manifest themselves as a narrow invisible ribbon running for several yards down the length of the coil. This streak of contamination might be located towards the centre, or towards one of the edges, of the width of the coil. Wherever it is located, it will render the cans made from it highly prone to splitting. It is for this reason that the total width of the coil is sampled (see Figure 4.4).

STATISTICAL CONSIDERATIONS

If there is no contamination of the coil by non-metallic inclusions, the conic fracture values of cans taken across the width of the coil will conform to a normal distribution, whose mean value is high and whose standard deviation is low.

Example 1   conic fracture values from an 'OK' coil:

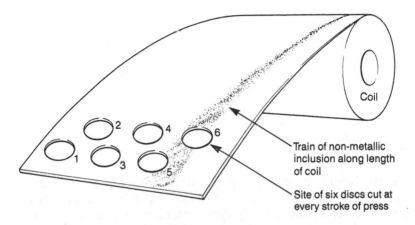

Coil

Train of non-metallic
inclusion along length
of coil

Site of six discs cut at
every stroke of press

**FIGURE 4.4   ARRANGEMENT OF DISCS CUT FROM COIL,
SHOWING CONTAMINATION BY TRAIN OF
UNCLEAN STEEL**

Disc No.	Value KgF
1	263
2	271
3	259
4	270
5	265
6	269

Mean = 266
SD = 5

These figures indicate a clean steel of consistent internal quality.

If, however, non-metallic inclusions are present in the coil as a streak of contamination:

Example 2   Localized substandard quality, fracture values:

These figures indicate a coil of clean high quality contaminated in the region of discs 5 and 6 with a streak of dirty steel.

The statistical assumption on which this type of calculation is based is mathematically incorrect. To compute all six of these values as if they belong to the same population, where they

Disc No.	Value KgF
1	282
2	260
3	273
4	267
5	176
6	159

Mean = 236
SD = 54

clearly belong to discrete populations (discs 1, 2, 3 and 4 comprising one population, and discs 5 and 6 being representative of another, significantly different, distribution) is to *inflate* the apparent standard deviation. Though this procedure is incorrect it is not invalid – the inflation of the standard deviation serves to emphasize the difference between the acceptable and the non-acceptable values of conic fracture. As such it is a tool of pragmatism rather than pedantry, which whilst perhaps offensive to the purist nonetheless serves the purpose of the practical.

And finally . . .

Example 3   Conic fracture values from a 'not-OK' coil:

Disc No.	Value KgF
1	193
2	206
3	189
4	187
5	166
6	178

Mean = 186
SD = 14

These figures, though of low standard deviation, indicate a steel whose low mean value confirms poor quality overall.

The tinplate suppliers accepted the validity of the conic fracture test and the usefulness of the data derived from it. They acknowledged the fact that the technology of making cans by the two-piece method places heavier demands on steel quality than does the three-piece system of production. Meeting these more rigorous requirements called for improvements in *their* technology – changing the metallic microstructure and its physical attributes and so on.

So, in adopting an uncompromising Right First Time approach to its own manufacturing operation a can making company not only solved a pressing production problem, it brought a positive and constructive pressure to bear on the suppliers of its main raw materials to do likewise.

This is Systematic Quality Control in effective action, keeping production costs down by minimizing wastage, all along the production stream, making a better product from a better raw material, using resources with a frugality which is of immediate benefit to the manufacturers and of ultimate benefit to the society which entrusts these resources to their stewardship.

Concepts of Folly and Futility: The Poisoned River

This is a story about a tribe of fools. They were wandering through the wastelands looking for somewhere to live, and they came upon a river of sweet rippling water. Being clever fools they built themselves a village on the banks of the river, and installed plumbing in all the houses, drawing fresh water from an uptake in the river. Being clever, *hygienic* fools, they also installed sanitary plumbing in their dwellings, building a system of sewers to carry away domestic waste which was ultimately collected into one main pipe which discharged its contents into the river, *half a mile upstream of the village's fresh water intake*. Then they complained every day about the taste of their early morning tea.

(Just in case you think this story is nothing more than a metaphorical example, which of course, it is, and that nobody in the real world would ever do such a stupid thing, recall the city of Hamburg in the late nineteenth century. The city fathers, faced with the problem of disposing of the city's human waste, built a sewerage system which discharged its contents into the river half a mile upstream of the city's fresh water supply uptake. In October 1890 over ten thousand Hamburg citizens perished from the poisoned river.)

We do something similar every day in our manufacturing systems; we poison our rivers of output by making far too many duds at the beginning of the process, getting it wrong First Time, and then try to put it right by pulling out the junk by depending on mass-inspection.

The Idiot's Tale: 4

Any multistage operation is vulnerable to pollution at source, and some kind of perverse imperative – a Murphy's law – seems to come into play to ensure that if a process is vulnerable to upstream poisoning then somebody will pour the poison in. The 'Rule' seems to be 'If it *can* happen, it *will* happen'. It becomes one of those familiar self-fulfilling prophecies, and more often than not it does happen. It did at Bedlam. What is more, if it *will* happen once it will usually happen *again*. It did this, too, at Bedlam.

Being a reasonable and logical sort of person you would think that once Bedlam had got it right with NOVA Cosmetics, they would firstly keep it right, and secondly extend this highly efficient Right First Time approach into their other customers' product lines. You would be wrong.

My job, as full-time priest for ever of Melchizedek and part-time quality manager at Bedlam, was to make myself redundant. There is nothing wrong in lifting crippled organizations off the floor, that is nothing more than your Christian charity and duty. There is plenty wrong with allowing yourself to be appointed to the position of permanent crutch. For one thing it ties you to one persistently ailing client and thus limits your scope to help others of the handicapped. For another this sort of dependence breeds yet more dependence, you spend your time propping up a sagging organization dedicating itself to its crippledom. This is no way to pass the days; anyway, it grows boring and it's very bad for the clients. So I declined to do it.

Hard on the heels of Bedlam's success with NOVA, and the consequent improvement to profitability, came the promotions. Little Boss was rewarded with wider authority, and given a whole new group of subordinates. In the prevailing compensation scheme of 'more heads equals more points, more points equals more pounds' it also fattened his wage packet. The Production Manager, who had watched the performance transformation with extreme hostility, was promoted to Factory Manager as a reward for simply being on the premises while it was happening, whereupon he terminated the high performance with extreme prejudice. This is a euphemism. In plain words he killed it. The 'key accounts executive' (another euphemism, it means 'salesperson') dealing with the NOVA account achieved his ambition to a title and became Sales Manager with a nameplate on his office door and a sunroof in his company car. 'What about me?' you ask. I asked, 'What about me?' The Boss asked back, 'What about you?' and that was the end of the performance appraisal and don't get any big ideas about being promoted.

It took about a year for them to undo all the good it had taken a previous year to do. Then things were back to the old status quo. Make it wrong. Sort it. Get it back. Sort it. Make some more. Sort it. Send it . . . back into the managerial maelstrom, the whirlpool of waste. Familiar waters.

Things could not go on like this, *again*. NOVA Cosmetics had by now given up all hope of getting sustained high quality from Bedlam and had shifted all their business elsewhere ('Thank heaven those pernickety nit-picking people have taken their damned business elsewhere,' said the Factory Manager with relief. 'Who needs customers like that with their endless going on about quality?'). Their place was taken by another equally demanding customer. This customer was not buying powder compacts, he was buying food pots, many thousands of little pots designed to contain convenience foods, and each pot printed to a design extolling the nutritional excellence of its contents. If cosmetic packaging is designed to offer the consumer the hope of beauty, food pots are intended to convey a concept of wholesomenes, of promise to promote fitness and health in the fortunate eater. The message of the container decoration, as well as meeting its legislative requirements of ingredient list and minimum fill weight, is also partly subliminal in as much as what is printed on the outside should project the cleanliness and trustworthiness of what's in the inside. The decoration is *important*. Quality of printing is vital.

The finished pots are made in two stages. First, they are injection-moulded on complex machinery from the finest food-grade polymer available. This is expensive. Second, they are printed on highspeed offset lithography machines with ink of the most glowing hues and which is guranteed non-toxic. This also is expensive. With such excellent raw materials and such sophisticated production processes it is not too difficult to produce thousand after thousand of containers perfect to behold, apart from the very rare and occasional pot – perhaps five or six in a thousand – whose decorative purity is defiled by some small aberration besmirching the printed surface. The customer was familiar

with the nature of this pot-production process and showed
extreme reasonableness by being prepared to accept any
batch of several thousand pots as long as it did not contain
more than one per cent of such blemished defectives. As
was said, it is not too difficult to achieve this.

Unless you are Bedlam plc.

If you are Bedlam, you pollute the river of output at its
source. You do this by splattering droplets of lubricating
oil onto the white plastic pots as they emerge from the
moulding machine. Not all of them – that would be too
gross and unrefined a way to pollute the river, just a few,
maybe two or three per cent. Not big drops of oil which
might be visible on the white surface even though the oil
itself is near transparent, *small* drops, *little* smears. Contamination which is not only difficult to detect at this stage,
but pollution which persists until the next – printing –
stage. At this downstream stage these light smears of oil
wreak their havoc. Oil resists printing ink. What is worse,
oil spreads. So every oily pot transfers its pollution to the
printing blanket as it passes through the printer head, this
oil smear prevents the blanket picking up its film of ink, so
when it transfers its ink to the pot it leaves a white uninked
smudge about as big as a thumb-print in the middle of the
decorative design. Then it does it to the next pot, because
the oil contamination from the original oiled pot persists
on the blanket for the next five or six passes, gradually
diminishing in area until after a while it disappears, and the
process resumes the printing of pristine quality again. This
is happening intermittently at the rate of one pot per
second. Too fast to be detected even by the trained but
fallible human eye.

This is how you do it, if you work for Bedlam.

The Production Manager contaminates, randomly
(which is to say not uniformly) with machine oil, about five
per cent of his output.

The Printing Manager, unable to screen out the offending pots, prints the lot. Five per cent defective grows into
twenty-five to thirty per cent defective.

Quality Assurance sample-inspect the output to British
Standard 6001 General Inspection Level II, to an Acceptable Quality Level of one per cent as agreed with the
customer. They reject more than half the submitted
batches, and mark them with dayglo red quarantine tick-
ets.

The Factory Manager, who thinks he is there to make
decisions, looks at the ticketed ranks of quarantined output and makes a decision. 'Send 'em,' he says.

The batches are loaded onto curtained trailers to keep

them clean and sent on their long haul down the motorway to the customer's factory.

The customer's Goods Inwards samples the batches and makes its decision. 'Send 'em back.'

The output makes the weary return trip up the motorway, home again.

The Factory Manager, who still thinks he is there to make decisions, looks at the jet-lagged junk and makes a decision. 'Send 'em to Madron Street,' he decrees. Madron Street is the School for the Mentally Subnormal in the street of that name just a few miles downtown. This is a therapeutic establishment with whom Bedlam have a long-standing business arrangement. It is a pool of cheap labour.

In Madron Street, amid the incessant chatter and happy cooing of those poor creatures of whom folklore says 'their heads have been touched by God' (why blame Him, then?) the junked output is 'sorted'. This is called work therapy. As if the afflicted inmates were not already burdened enough without this additional imposition of senseless and unnecessary drudgery. The sorted output is returned to Bedlam, relabelled and sent down the motorway yet again.

By now the customer is desperate for some packaging material, so he returns only twenty per cent of deliveries.

The Factory Manager, unshaken in his conviction that he exists to make decisions, decides . . . 'make some more to cover the wastage.'

And the whole unhappy cycle goes round and round in Bedlam's bottle-to-bottle hook up.

They are not running a business. They are running a charity. A charity whose four beneficiaries are:

The Raw Material suppliers. They are more than pleased to supply Bedlam with twice as much material as they need because half of what they buy they waste. Bedlam's centralized Purchasing Manager has pared the last penny off the buying price, he's done his bit, what Production do with the material when they get it is none of his business. He might just as well not have bothered.

The Scrap Merchant. Half of the material Bedlam buys eventually finds its way to him, at one tenth of the cost Bedlam bought it for. His is happy. His *second* car is a Rolls Royce. He is one of the fat vultures scavenging the landscapes of manufacturing industry, gorging and growing rich on the carrion of their incompetence. He is a generous man, and provides about half of Bedlam's management's Christmas drink requirements, as a token of his esteem. He can afford it.

The Trucking Company. Their job is hauling things up

and down the land. They are paid mileage. Whether they are trucking OK output or junk is all one to them. Back and forth, round and round, and every mile paid for by Bedlam. They too are generous; they supply the other half of Bedlam's Yuletide beverage needs, to express their happy wonderment at their client's folly.

The School for the Mentally Subnormal. They have no objection to going through the lucrative ritual of 'sorting' dud output from rejected batches. It provides at least the illusion of meaningful work to innocent souls whose lives seem to hold so little in the way of meaning. But who knows?

The customer was sick of it, and had started to look elsewhere for a supplier to replace Bedlam. He phoned Big Boss and advised him of this serious turn of events. Big Boss had to do something, if only to deflect the customer from contacting Head Office and ripping a hole in that screen of deceit he called his Monthly Report, through which a bit of the sorry truth might be discerned. We can't have *that* can we? So he summoned his Wandering Maverick messiah from one of the other regions of the empire. 'Trouble in Plastics with food pots,' he told him. 'Get it sorted.'

'Again?' I asked. 'I've done it once.'

'Do it over again then, that's what you're *for* isn't it?'

Nice to know what you are for. When a child points at you and asks 'What is that man *for*, Daddy?', as the little girl enquired about a famous politician, it's good to be able to give a sensible and worthwhile answer, which Daddy was unable to do in response to his daughter's question. I was there to 'get it sorted'.

'With your full authority and support to do whatever is necessary to bring things back onto a first-rate footing with the customer?' I asked – always good to know how much muscle you can pull.

'Of course,' he snapped. That is one of the things I had grown to like about Big Boss: once your ability had earned his respect it brought his fullest support. No matter how you did it, as long as you could do it.

So I did it. After all, that is what I was for.

How?

Surely the answer is obvious. Simply enforce the system already in place but being bucked by misapplied authority. Stop the pollution of the river at its source. The details of how this was achieved will be described later in the text. This called for the exercise of considerable power. We shall be discussing the three-pronged nature of power later in the text; an understanding of it is important to those who would change things. As part of this examination we shall look at mythology and myth-making, and how it worked so wonderfully in this instance that in no time at all Bedlam climbed from the most inferior position on their customer's vendor assessment ranking order to become Number One. Again.

Summary

Deming's Third Point demolishes one of the myths of old-order management which has been a keystone of manufacturing activity since the beginning – *inspection*. Inspection of the product *after* the manufacturing event *cannot* be successful. It is too late, reactive, and divisive to the parts of the organization.

Quality seeks to control the product by controlling the process, thereby rendering final inspection of the product completely redundant. Quality becomes an *invisible input*. The manufacture of tinplate opentop food cans is a splendid example of the success of this approach.

This no longer subscribes to the old concept of folly and futility known as 'the poisoned river', where upstream defectives foul downstream operations and result in duds being shipped to the market.

Bedlam were poisoning their river. Management was working hard on *localized optimization* to the detriment of global optimization. Each was looking after his little part and nobody was seeing to the whole of the business transaction. The poisoning had to be stopped.

5 DEMING'S FOURTH POINT

> 4 End the practice of awarding business on the basis of price tag. Purchasing must be combined with design of product, manufacturing, and sales to work with the chosen suppliers: the aim is to minimize total cost, not merely initial cost.

Sometimes, when a couple is making war instead of love, you hear 'she is an awful wife', or 'he is an awful husband'. Wrong! Both statements are unfair and untrue. She is not an awful wife nor is he an awful husband, but they certainly have an awful marriage. It's the *relationship* which is defective, and it takes two to tango.

You hear similar verdicts in industry. 'They are awful suppliers', or 'So-and-so are lousy customers'. Wrong again! There is no such thing as an awful supplier or an awful customer, but there are some lousy relationships established between the two. It is the nature of their *mutual transaction* that matters, the quality of the relationship.

Relationship Styles

To use again some of the material published in the book *Right First Time* . . . The style of the relationship between a customer and a supplier can be described as being one of three kinds:

ADVERSARIAL

'The Customer is King.' It says so on the yellowing poster sellotaped to one of the walls in the General Office. This poster

74

has been hanging there for as long as anyone is able to remember; it depicts a lion, gazing with the vacuous majesty of a moronic monarch to stress the sovereignty of 'the customer'. Nobody gives it even a first glance any more, let alone a second. In any case, the people assembled in the office are too busy tackling problems – well, *talking* about tackling problems – to be bothered with the wallpaper. They are congregated for the ritual of the Daily Management Meeting. This is an ancient ceremony, its origins lost in the mists of managerial memory; the lion has borne impassive witness to hundreds of such gatherings down the long years of his ennui. Like any ritual the performance is repetitive, unchanging, and therefore reassuring in a changing world. Any valid meaning this meeting might once have had has long been lost, its only purpose now is the perpetuation of itself. Whatever valid meaning it might have had has been perverted into a finger-pointing exercise. It is a witch-hunt. It is faithfully re-enacted every working day at ten o'clock in the morning. The plot is unvarying, the theme is fixed, the cast constant; only the script varies, and that not much. A witch-hunt needs a witch. Whose turn will it be today?

Not the Technical Manager's. He's arrived ten minutes late with an excuse on his lips and grease on his hands to prove he's been busy persuading some recalcitrant machine to run how it's supposed to. He would have got it going sooner and been here on time if Quality hadn't interfered, with their questions of machine capability. He gets a 'well done' and smiles contentedly in the knowledge that he's safe for another day.

Not the Production Manager's. He stands within the secure protection of the unassailable barricades of his output figures. They are good figures, he knows, because he's flogged the machines to their limit to get them. He scowls aggressively over his battlements to show he's taking his job seriously, and lets it be known that the figures, good though they are, would have been even better if it handn't been for Quality interfering again, insisting on 'the specification'. He gets a nod of approval, which relaxes his scowl into a smirk.

Not the Sales Administrator's. All despatches have been made on the due date, no lateness. Except, that is, for those batches of output held in quarantine because Quality are being awkward again and talking about infringements to specification. Don't know why Production let them get away with it, *we* are the ones who have to deal with the complaints from customers about late delivery, and it's not our fault. You make it, we'll send it. We're on the ball we are. This gets another curt nod of approval, to permit Sales Administration to breathe more easily.

For some reason the Sales Manager has chosen to grace today's meeting with his presence. You knew he was on the site when you saw his car in the front park – his is the one with the shirt hanging in the back, evidently sales persons are able to choose a washing line as an optional extra when they specify their company cars. Wonder what he's doing here? Rejection, he says; a whole delivery rolling back home up the motorway, rejected for failure to meet the specification. What have Quality been playing at?

A witch-hunt needs a witch. *Whose* turn will it be today?

The *Quality Manager's*, you say? Well, I never did! Fancy that! You *do* surprise me.

The Quality Manager is given his instructions, his royal command, you might say his bit of papal bull. The returned delivery will be turned around as soon as it arrives and sent out again to the complaining customer. So will the Quality Manager. He will share the Sales Manager's car with the Sales Manager's shirt and be conveyed to the same destination as the offending load. Once there he will convince the customer that there is nothing wrong with the delivery. Failing that he will make an act of atonement. The customer will be wined, the customer will be dined, the Quality Manager will sustain himself on a diet of humble pie. He will not, repeat not, let that delivery be rejected. Above all he will never let any credit note be authorized, as these are visible at Head Office and we don't want *them* to find out, do we? In the last extremity he is empowered to offer to replace the entire delivery free of charge, but this will be taken as a sign of his failure to manage the situation and control the nuisance of a customer.

The lion looks down. The Customer is King.

The philosophy underlying this style of supplier/customer relationship is price-tag purchasing. The buyer sets out to buy at the cheapest possible price. *There is nothing wrong with that*, as long as . . .

If the buyer trawls the supplies market in search of the lowest possible price, *regardless of all other considerations*, this is the kind of relationship he is, perhaps unwittingly, fostering. Sooner or later it will lead to this confrontation. The aim of this kind of purchasing policy is to minimize *initial* cost and, as Dr Deming tells us, the aim should be to *minimize total* cost.

The majority of enterprises these days, both in the manufacturing and in the service sectors, buy in more than they make in. The Purchasing function is a Big Spender. Its decisions affect operational performance more than they have ever done in the past. There is no place any more in this bigger scheme of things for Adversarial-style relationships. So what other styles are there?

INTRUSIVE

Quality can sometimes be a matter of life or death, victory or defeat.

> For want of a nail the shoe was lost
> For want of a shoe the horse was lost
> For want of a horse the rider was lost
> For want of a rider the battle was lost
> For want of a battle the Kingdom was lost
> And all for the want of a horseshoe nail.

Perhaps the nails were made of cheap and brittle iron and they snapped too easily; perhaps the Supplies Officer who bought them for his regiment of cavalry congratulated himself on his business acumen for having screwed his nail-supplier down to the last contractual penny. The nail-maker, still needing his modicum of profit, used cheaper strap-iron to forge the nails than he would otherwise have chosen, iron contaminated with too much non-metallic slag. Perhaps when the farriers were shoeing the horses the nails were snapping like cinnamon sticks, so the smiths blamed the supplier. Perhaps when the cavalryman's horse cast a shoe he blamed it on the farrier. The defeated general knew nothing of all this trouble right back up the supply chain. So he could hardly blame the nail; such a mighty contest lost for such a trivial thing?

The military have always been alert to the danger of accepting substandard supplies; when your life might depend on the quality of what you have purchased it must focus your mind wonderfully on 'quality'. The Procurement Directorate of the Armed Services learned very early on in the game that a good way of ensuring the quality of sup*plies* was to ensure themselves first of the quality of the sup*pliers*. So they developed the *intrusive* technique. This involves looking at the potential supplier's *system* of controlling the quality of his output, and eventually led, through Defence Standards, to British Standard 5750 and to the ISO 9000 series.

These are systems of quality management. They intrude by asking the potential supplier such questions as: 'Who in your organization looks after quality?' 'How well qualified is the Quality Manager?' 'What power does he (or she) wield in the hierarchy?' 'Who does she (or he) head into?'

'Are records kept of this ... that ... the other?' Over a hundred questions whose answers are intended to assure them that the potential supplier is operating an effective *system of quality management*. A system is essential. But a system is only a system, nothing more. A system cannot build quality into the product any more than scaffolding can build a pile of bricks into a house. It is a framework, within which work is done. It is all structure, all Form, and no Content; a Church without a Christ. BS 5750 is such a system. Many companies nowadays require their suppliers to be certificated to this standard. Implicit in the imposition of this standard upon suppliers is the message: 'It is not our business to know or care whether or not you produce poor quality output, our only concern is that *none of it gets to us.*' BS 5750 does nothing for preventive quality.

There are companies who have achieved accreditation to BS 5750 with the sole purpose of being able to boast about the fact and use it as a competitive edge. Fair enough, there is no harm in that. But some of these companies still experience horrendous quality problems in-plant, which is a shame, as it misses the spirit of BS 5750 through pursuing the letter of it.

A number of manufacturing organizations have their own equivalent of BS 5750. They require their suppliers to demonstrate quality competence within a system of quality management. So they intrude, they question, they assess, they select, they reject.

A not uncommon aim of companies using this approach is the reduction of their supplier-base, buying more and more from fewer and fewer *approved* suppliers. Suppliers whose fitness to continue supplying is regularly monitored and measured. If there is ever to be any confrontation in this style of relationship, it takes place long before any deliveries have been made. The essence of this philosophy, unlike the Adversarial, is concordant not confrontational.

This is a good style of relationship, but there is an even better one.

ENHANCING

The Western manager, on his obligatory 'fact-finding mission' to Japan and the Pacific Rim, was questioning an eminent Eastern industrialist. He asked him what the secrets of his success were, how he brought it about in the complex world of manufacturing, how he got so much good work out of his people, what it was he *actually did.*

The industrialist said 'First, we educate,' and described some of his company's training and educational programmes. What did he do after that? 'After that, we educate,' he was told. Then the Western manager enquired about the next stop. 'Next we educate' was the reply. At this point the Western manager experienced a strong sense of *déjà vu,* and hazarded, 'I know what you do next, next you educate.' The industrialist replied, 'No, we trust.' The manager was curious to know what happened after that. 'After that, we inform,' he was told. And after that? Succeed.

For a long time education has been a despised commodity in Western managerial thinking. To describe somebody as 'brainy' is to deliver a career-killing insult. Western managers pride themselves on being 'practical' people who hold learning as 'theoretical', or 'book-learning'; theirs is a predominantly artisan culture. To their loss. But times are changing as the world of work is changing. 'Education' is becoming a prerequisite of survival, in the not too distant future all companies will, willy-nilly, be teaching companies. Not only teaching their own people, but educating their suppliers as well. For example . . .

'Ford cares about Quality.' Education is the key to quality. So

Ford cares about education. There is an inescapable logic in this. If you are in the process of narrowing your supplier base from over three thousand to less than three hundred, the survivors had better be good – in fact they had better be the best. So the Ford Motor Company's approach to supplier quality goes much, much further than the Intrusive style (though they do most assuredly intrude, through the instrument of their Ford Q-101 system), they insist on *enhancement*. Good you may be, but to get on their list you had better be better. They devise educational programmes to teach Statistical Process Control to their suppliers, because they know that you can only get the best from the best, and they aim to be 'best in class'. Their purchasing policy is absolutely in line with Deming's Fourth Point, 'Purchasing must be combined with design of product, manufacturing and sales to work with the chosen suppliers.'

There is no other way.

More about Customer/Supplier Relationships

Japanese manufacturers are eminently successful. They domi-nate so many markets. Motor bikes, microwaves, ovens, watches, ships, television sets, and so on. But try this one. In any assembly of people ask all those who own something made by Black & Decker to raise their hands: just about everybody's arm will go up. Then ask all those in the congregation who own anything made by Makita to raise their hands: about one in thirty will go up. Makita is the Japanese equivalent, and presumably competi-tor, of Black & Decker. Why so few? Are the Japanese makers of small DIY electromechanical appliances less aggressive than their compatriots in other fields? Surely not. Is Makita's equip-ment either prohibitively expensive or of inadequate quality? Certainly not. Is it available in the shops. Of course it is. So how come Black & Decker maintain their position of market pre-eminence? The words of Roger H. Thomas, Black & Decker's Group Vice President Europe, might provide a clue:

> It was a deliberate policy to embark upon an on-going process of trying to do the best in all our divisions, to ensure that the customer gets a high quality product; that it represents value for money; and that he will buy another Black & Decker at the end of its useful life.
> Two years ago we embarked upon what we call a TCS culture

change. TCS stands for Total Customer Service, which means that everyone in our company, in everything he or she does, does so for a customer. The customer more often than not will be inside Black & Decker e.g. a gear is cut in the machine shop and goes to the assembly shop. It is important that the gear cutter knows who his customer is and identifies himself with his customer; understands his customer problems; and how he indeed can represent better value and provide better customer service. You can extend this to secretaries or, in reverse to bosses who dictate to their secretaries, and so on.

It is a culture change which will take many years to install if we are able to be successful.

Observe the eloquent simplicity of their philosophy of Total Customer Service. This is 'simple' in the very best sense of that sometimes misunderstood word. Here is a message of zero ambiguity. It means exactly what it says, and it says exactly what it means. It is a message – *not* a slogan – stripped of anything superfluous or irrelevant. Three words which say it all. It looks as if it works.

Now try another one. Ask one of your salesmen, 'Who is your customer?' He will stare blankly at you for a second or so and then recite the names of all the *companies* in his sales area. Then ask him, 'But who is your internal customer?' After his bemused pause help him out by asking 'Who in this organization do you pass your orders to, so that they can be processed into deliveries and invoices?' Now the penny drops. 'Oh,' he tells you airily, 'I pass them on to Yvonne in Sales Administration, she looks after that end of the operation.' So Yvonne is his 'customer'. When did he last go to her and ask how he was meeting her needs as his customer? It might be interesting to find out.

In terms of the Handy/Harrison cultural model this total customer service concept is well into the right-hand side of the horizon profile, well into Achieve (Task)/Support (People) culture. This surely validates the proposition that 'company performance is a function of company culture'? Or, to put it less pompously, 'you *do* what you *are*'. And seeing that you can be whatever kind of culture you want to be, it follows that you can do whatever it is you want to do. Even dominate a ferociously competitive market sector. The only person preventing you is you.

This is all good stuff, but, thanks to good old human nature, it has its downside; this is it . . .

Mankind seems to be possessed of an awful knack of reducing just about everything to the level of the banal. It seems to be to do with adopting some kind of mental shorthand to save having to think. For example, BS 5750 is an excellent system of quality management; Statistical Process Control (SPC) is perhaps the most powerful and versatile tool in the entire statistical kit. Yet no sooner do some people hear these acronyms than they endow them with a near-talismanic significance. They become charms, chants, amulets to ward off the evil eye. It is as if to merely utter 'BS 5750, SPC' ten thousand times is to begin to see the stars in the firmament going out one by one. This atavistic recourse to the magical finds its manifestation in business when the Purchasing Manager, who has lately heard of these things, asks of a potential supplier, 'Do they have BS 5750, do they do SPC?' and when told yes says 'That's OK, then' and places his contract. To be effective in his newly-enlarging role the Purchasing Manager should cultivate a deep insight into what these things are really about, then go and meet the vendors who are supposed to 'have' BS 5750 and are 'doing' SPC to find out for himself through informed enquiry. After all, business is transacted between *people*, and more often than not between *friends*. What kind of people? What kind of friends?

The Japanese said, 'In your country you say customer king, in my country we say customer God.' Well . . . yes . . . but does that mean that the supplier must make obeisance to, must *worship*, the customer? What kind of one-way transaction do you call that? That is not co-operation, it's conquest. There is another way of arranging things.

This chapter began with a mention of that transaction between two parties known as 'marriage'. Let us conclude this section of it on the same topic. First of all we shall quote Abraham Maslow on the subject, and then paraphrase his passage to suit our business context.

Therefore strong men and strong women are the condition of each other, for neither can exist without the other. They are also the cause of the other, because women grow men and men grow women. And finally of course they are the reward of each other. If you are a good enough man, that's the kind of woman you'll get and that's the kind of woman you deserve. (Abraham Maslow, *The Farther Reaches of Human Nature*, p. 94, 'Creativeness'.)

Now let us substitute 'supplier/s' for man/men, and 'customer/s' for woman/women . . .

'Therefore strong suppliers and strong customers are the condition of each other, for neither can exist without the other. They are also the cause of the other, because customers grow suppliers and suppliers grow customers. And finally of course they are the reward of each other. If you are a good enough supplier, that's the kind of customer you'll get and that's the kind of customer you deserve.' Which seems to sum up the essence of the Enhancing style of supplier/customer relationship. One of mutuality. One of *quality*.

You have to work at it. Together.

Purchasing Criteria

WHY BUY?

There are three factors involved in Purchasing; they are Price, Service and Quality. Which of these is, to use the jargon of the trade, the Primary Determinant?

A piece of research carried out by Barrie Dale and Peter Shaw at UMIST conveniently confirms what was already suspected, namely that the most potent factor affecting purchasing decision-making used to be price. Followed by service and trailed by quality. Until a few years ago. The sequence has switched, it now goes quality, service, price.

BUY WHAT?

Products are bought to *specification*. What *is* a specification? Never mind the marvellous and worldly-wise definition put forward by Alan Cowan three decades and more ago, 'a specification is a document written with a pen dipped in tears'. Oh dear, and how, so often, true. Let us look at a hypothetical specification . . . Suppose you are a manufacturer who makes things out of sheet metal. I am a buyer who wants several thousand metal discs. Here is my spec.:

1 The discs shall be made from metal Grade A with the

QUALITY = Exact Fitness for
Intended Purpose, Achieved
at Economic Cost.

PRICE — of Minor and Dimishing Importance

SERVICE — of Secondary Influence

QUALITY — of Fundamental Influence, Primary Purchasing Determinant

FIGURE 5.1   THE NEW PHILOSOPHY SEEN FROM THE
VIEWPOINT OF MARKETING STRATEGY –
QUALITY AS COMPETITIVE EDGE

properties P1, P2 etc. of thickness T. (Assume you have
enough of this grade and thickness of metal to hand.)
2   The disc diameter shall be 1.300 inches, with a tolerance of
plus and minus .005 inches.
3   The discs shall be round.
4   They will be a good fit in a hole whose diameter is 1.306
inches.

What do you think of *that* as a specification? Hypothetical it
may be, but there are plenty like it about the place, please be
assured. Do you, as a potential supplier, accept it?

Before you answer, consider this – you go to a cartographer
and ask him, 'What is the waterline distance around the British
Isles, excluding the Isle of Man, the Isle of Wight, or any other
of the islets dotted around the coast?'

He replies, 'Do you wish me to make this measurement at High
Water Springs or at High Water Neaps, or do you wish me to
measure this wet distance at Low Water ditto? Or do you want me
to measure the distance at half-tide? Do you wish me to include
the wetted distance around every boulder or every pebble in my
total measurement? In answer to your question I can only say I
don't know, and neither does anybody else, and nobody ever will.'

Then he asks you why you want to know this distance. You
tell him that you plan to circumnavigate the British Isles in
your boat.

'Ah,' he nods, 'then all you need is the waterline distance around the land as measured from one convenient headland to the next. This information I am able to give you, because yours is now an *operational* specification.'

So how about the specification for the metal disc, is it 'operational'?

What does 'round' mean? What does 'good fit' mean? Not much.

This 'specification' is far too ambiguous to be of any use, transact business on this document and when the crunch time of acceptability or rejectability of batches arrives the problem will never be resolved.

How round is 'round'? This can be expressed numerically as a statement of acceptable ovality.

What is a 'good fit'? Forget it, nobody will ever agree on it. Strike it out of the spec.

Specification writing can be a tricky business, to do it wrong sometimes ends up in the writing of credit notes, with a pen dipped in tears. Yours.

The Idiot's Tale: 5
Here I am again, Aaron Godman the Enigma, Quality Man and your Guide and Mentor in the Madhouse of your manufacturing world. Welcome back to Bedlam. Not into the factory which makes things out of plastics, this time into a sister factory where things are made out of sheet tinplate; but for all managerial purposes just a different ward in the same asylum. With similar problems.

Problems are the lifeblood of the incompetent manager. Problems are so much more fun than solutions. Solutions are boring, they are about doing things with a competence so calm it seems almost casual, and therefore earns the disapproval of those bosses who recognize activity but are blind to achievement, like Bedlam's. Besides, when you have problems you can shout at people, that's fun too.

These bosses believed themselves to be 'astute businessmen'. Astute businessmen buy cheap and sell dear, so they bought cheap, and this minimized initial costs. But unaware of the good sense of Deming's Fourth Point, in minimizing initial costs they fell into the trap of maximizing total costs. Maximizing to such an extent that every penny of profit disappeared down the plug-hole of their ineptitude. This is how they did it . . .

They have been awarded a contract by the well-known

High Street retailing giant Engels and Frank. A contract to supply many thousands of powder compacts, made as decorated tinplate boxes with a hinged lid, the lid colourfully printed to depict the profile of a beautiful woman. The artist's design, to which they were required to work, endowed this profile with a complexion of flawless angelic beauty, a Nefertiti-like perfection in glowing pinks and translucent creams. No problem, the technology of printing on tinplate is well advanced and quite able to cope with artistic demands such as this. You simply pass sheets of tinplate through a printing machine, and on each sheet you print in many colours a number of identical impressions, each of which will later become an individual lid. There is one slight snag; any surface imperfections on the tinplate will show through the thin coating of printer's ink which is applied to the sheet.

These were astute businessmen, who knew the tricks of buying cheap and selling dear. They hadn't got where they were today by not being astute; business acumen, that's what they had got. So they purchased a large quantity of tinplate sheets of the correct material grade and thickness, cut to the correct size to be able to maximize the number of printed impressions per sheet without material wastage. All praiseworthy production planning. They ordered a few extra to allow for a bit of scrap. Tinplate mills are not in the 'perfection' business, they sometimes produce substandard output. They will knowingly, and with your agreement, sell you batches of tinplate most of which are of a prime quality but contain a proportion – say 10 per cent – of sheets which are known as 'unmended menders' (presumably because they are wrong and therefore need 'mending', but have not been 'mended' so are 'unmended menders'. All very archaic and esoteric and quite logical; I suppose it's as good a way as any of describing junk output.) If you are prepared to pay for prime plate, the tinplate mill will pull out 'all' the unmended menders in the batches they deliver to you, which costs you extra. For a lesser price they will sell you standard unsorted output containing substandard sheets. For least of all they will sell you unmended menders sifted from the batches they have sold to other people as prime quality.

Bedlam costed the selling price of the tin boxes they were going to supply to Engels and Frank on the basis of using prime quality tinplate. Then they 'minimized initial cost' by buying unsorted output at about 15 per cent cheaper than their budgeted raw material cost. More 'business acumen'. Smart work this, 'fifteen per cent straight to the bottom line before we've struck a blow', they boasted to themselves; 'this is how to manage a business'.

Is it? This is *buying trouble.* You want a problem? Then buy one in! Buy second-rate raw material. Go on, buy it, it's cheaper isn't it? You're a businessman, aren't you?

Or are you a blind and greedy fool? Look at the consequences of your 'astute' decision with these powder compact boxes. The ink used to print designs on tinplate is not opaque, and it is applied as a very thin coating. Any surface imperfections on the tinplate will be visible through the ink after printing. The whole point of painting the profile of a beautiful woman with a complexion compounded of rose-petals and the cream from molten gold is to sell the hope of beauty to the girl over the counter. But when Bedlam printed, with great skill, the substandard sheets, the complexion of the profiled girl came out patchy, to say the least. Every surface defect on the tinplate showed through the ink. On some impressions this gave her the aspect of a serious case of acne; on others her splendid cheekbone seemed to have been whipped with a bundle of stinging nettles into an angry rash. She had blackheads. Her face had been raked with briars. Sometimes she appeared with leprous blotches disfiguring her lustrous skin. She occasionally sported what looked like a Heidelberg duelling scar. She had even developed a stubbly androgynous seven o'clock shadow. Not really the marketing image which Engels and I rank wished to project.

Sometimes she looked OK, as the artist had intended – about seven out of ten impressions were flawless. Bedlam decided to sort out the duds, and thus jumped into that whirlpool of waste which is the hidden cost of poor quality. Having paid the printing costs, they added the sorting costs. You know how effective 100 per cent sorting is? Then they paid transport costs to ship the salvaged material to the customer. Then they paid it again to ship it back. Then they paid some more sorting costs. Then they bought some more tinplate sheets to print over again to make up the shortfall caused by the high rate of first-time failure, then they . . . pay, pay, pay! And keep on paying, this is the cost of managerial 'minimization' of initial cost, the penalty of greed and price-tag purchasing; and if this is not insane behaviour appropriate to a madhouse then nothing is. They should have paid heed to Deming's Fourth Point.

Summary

Deming's Fourth Point tells us that purchasing by price-tag cannot work. It is too narrow an approach to the transaction between supplier and customer, too limiting, too fraught with too much risk of getting it wrong. It actually institutionalizes the purchasing of trouble.

Of the three styles of relationship between suppliers and customers – the Adversarial, the Intrusive, and the Enhancing – only the third style has any chance of achieving long-term and consistent success.

Systems of quality management, their benefits and their limitations. The importance of an *operational* specification, and of not buying trouble when you buy raw material. If you begin with good quality you may end with it, if you start with poor quality it will never be possible to screen it out.

Bedlam discovered this, again, for the umpteenth time in their history, it can be a terrible and costly lesson.

6 DEMING'S FIFTH POINT

5 Improve constantly and forever every activity in the company, to improve quality and productivity and thus constantly decrease costs.

A certain delicacy attaches itself to this point: to suggest improvement is to imply inadequacy. To say things could be better is to state that things are not as good as they might be. To recommend change is to criticize implicitly the status quo. The architects of the status quo might not be too pleased with this, nobody likes being criticized, either for himself or for his works. This is a quite understandable touchiness and it sometimes poses a very real problem to the would-be change agent. To initiate change often calls for a high degree of tact on the part of the change-agent; and for a degree of humility and willingness to accept the need for change on the part of the person whose operational practices are about to be the subject of controlled and deliberate change.

Sometimes the second condition does not prevail; if the person who calls upon the services of the change-agent does so because they are already dissatisfied with the way things are being done, the prospect of change offers not threat but opportunity. Opportunity to *improve* something. Their outfit is more than ready to undergo change, if only it knew what kind and direction of change it wanted or needed.

Why bother? Why indeed. Some companies don't. They're doing nicely, thank you, turning in a tidy little profit, nothing spectacular you understand, but enough to keep the wolf away from the door. This is the complacency which foreshadows the catastrophic shock of collapse, or the trauma of conquest by a predatory take-over.

However, in the majority of enterprises, and especially so in the higher performers, there exists no opposition at all to Deming's Fifth Point. In these organizations 'continuous improvement' is seen to be one of the honing-stones used to whet the competitive edge. This is another of the many paradoxes which dwell in that Pandora's Box of paradoxes called 'management of change'; higher performing companies seem to say to themselves of themselves, 'OK, we're pretty good at what we do, we're holding our own with the competition by the look of it, so *how can we become better?*' In contrast there are the mediocre performers who echo this opinion of themselves, but omit the 'how can we become better?' bit of it. So they make no progress, which does not mean they stay where they are, relative to the competition they *fall behind.* They see themselves as successful, and forget that nothing fails like success.

So how do we begin to apply the precepts of this Fifth Point? Where do we start? *Everywhere*, in every activity? This is what Deming suggests but it seems a tall order, with some danger of biting off more than can be chewed. Perhaps we should not be too ambitious to begin with; we shall certainly aim to fulfil the whole of Point Five *eventually*, but in the meantime settle for a more modest target. Does that sound like a worthwhile approach? Let us take a look at a case study from a consultant's assignments journal, to see how this approach was used to good effect.

The Black Art

The client company makes things out of black rubber; things like huge gaskets for filtration plants, anti-ripple tiles to clad nuclear submarines, shock absorber pads for the automotive industry, track links for armoured fighting vehicles . . . and a host of other components where the special properties of rubber are required. They were, and are, a successful organization, but their MD knew that though they were doing well they could do better; so he called in a change-agent for a chat.

The two of them were strolling through the plant, passing through veritable canyons of black rubber, threading their way between islands of black rubber, skirting mounds of black rubber. 'You know,' said the MD, 'there's a touch of the Black Art about what we do.' The consultant thought this was some kind of pun on the rubbery blackness of the landscape, and nodded, 'I think I see what you mean.' But the MD continued, 'We have certain production problems, which are costing us a lot of money. My Technical Director tells me we have a delinquent process.' There it was again! The old familiar anthropomorphic attribution of human mischievous intent to an inanimate process. The consultant had heard this one umpteen times before. He had likewise heard the next of the MD's observations. 'The Chief Chemist says it's all the fault of the raw material.' Well, it would be, wouldn't it? These are from that litany of cop-outs used by the baffled to 'explain' the inexplicable. We are not doing so well, it *cannot* be our fault, therefore it must be the fault of ... process, raw materials, things have always been this way ... OK why not say it is the fault of excessive sunspot activity, or the curse of the pharaohs, or just plain bad luck? As 'explanations' of failure to perform these latter are no less valid than those customarily cited. Why not the truth? Plain incompetence. Good old-fashioned perplexity with the day-to-day phenomena we meet in manufacturing but are unable to account for in meaningful terms because we don't know the *questions*!

The MD went on, 'A few months ago we called in a team of academics from the local polytechnic to teach us all about quality. It doesn't seem to have done us much good. Everybody over the level of supervisor was exposed to this programme, yet still we are making too much defective output.' The polytechnic mentioned is acknowledged to be among the best teachers of quality philosophy and practice in the country, so why had the training programme had so little effect on their performance?

'Because you don't know the right questions,' advised the consultant. The training in quality has provided them with the *means of answering* their quality questions, but they were unable to ask the correct 'why' about the problems they saw. This is what quality is all about. It is *not* about pulling masses of data out of the process willy-nilly and analysing it. Any kid with a calculator can 'analyse'. The trick of quality is *knowing what the question is.*

What was the question in this establishment? The process was misbehaving. The raw material was acting awkward. These were the 'reasons' for excessive variability in the product. There is a third factor in the process equation as well as material and machine – there is the operator. Why was *he* being left out of their reasoning? Because he was *only* an operator, with nothing to contribute to the solving of the problem? To be fair, if you have been part of the problem for nigh on fifteen years, as was the case here, it's not easy to become part of the solution. This calls for a change of polarity in the mindset. So where and how do we begin, what are the steps, do we try and do all at once the things in Deming's Fifth Point, to 'improve constantly . . . every activity . . .'? No. Such instant and total success is not what Deming means. We do it the way the tide comes in, one wave at a time. So the first wave of the chosen project laps up the beach, and wets the sand that has been dry for long enough. Then it recedes a little but not as far back as its start point. The second wave finds it easier and swills higher up the shoreline before it crackles back a little way down the shingle. Wave by wave the sea comes in, until the highwater mark of knowledgeable competence is wetted. In a word, there is a *procedure* for tackling intractable problems, and it's as powerful in shifting the obstacles of solid unknowing that are a barrier to achievement as the sea is in shouldering away the boulders which stand stolid in its path. This is it:

1 Select the *right* project. Tackle one problem only (to begin with).
2 Give it *high* visibility, take something that's been a problem for so long that common knowledge says it will always be a problem, as if it were the decree of some spiteful divinity.
3 Give it high credibility, in the only terms which are readily understandable by everybody and refutable by nobody, namely *money*. Put a price tag on success.
4 Form a trinity of alliance, then
5 Do it.

The 'right' problem is one which your experience, as a peripatetic witch doctor, tells you is the most amenable to being solved. This is a basic precept of generalship, to attack the enemy where he is weakest, to overwhelm, and to ignore the rest of the battle front. The selection of the target area is purely a matter of judgement, and to judge wrongly is to lose.

There happened to be one obvious target in the battle front of problems facing the rubber converters. To appreciate how *simply* it was sorted out it is necessary to look a little into the technology of rubber processing, after which we shall go through the four steps of the corrective procedure.

THE PROCESS

Rubber is manufactured as a batch rather than as a continuous process. It is compounded from latex, a naturally occurring resin, sulphur to help it vulcanize, carbon black, chemicals to endow it with desired properties and other materials in a mixing hopper. From here it is delivered to a 'mill', which is really a giant wringer in which the rubber is squeezed into sheet form by passing it through two large steel rolls, each about six feet in length and about two feet in diameter. As the solidified sheet, about one inch thick, emerges from the rolls it is cut into a length and width convenient for handling. The mill operator produces what are to all intents and purposes 'dartboard mats' about seven feet long and about two feet wide, but still about one inch thick. These are rolled, much as you would roll a stair carpet, into handleable 'slugs' suitable for insertion into the machine which performs the next stage of the production process. This stage is 'extrusion'. Three or four slugs are inserted into the barrel of an extrusion machine, whose breech is then closed and clamped securely to contain the extruder contents. See Figure 6.1.

Heat is applied to render the solid rubber into a viscous state which will enable it to flow under applied pressure. Pressure is applied to the rubber in the rear of the extruder forcing it through a shaped die at the opposite end. As the stream of rubber passes through this die it assumes the cross-sectional shape of the die, and cools and resolidifies. The extruded rubber continues to emerge from the die in an unbroken stream. At this point the extrusion is cropped into a succession of separate

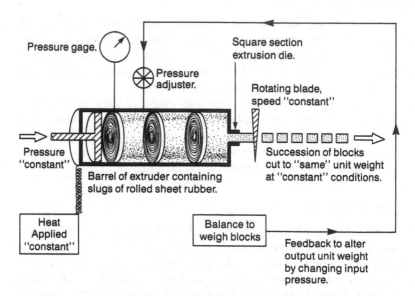

FIGURE 6.1   EXTRUDING RUBBER BLOCKS TO 'CONTROLLED'
             UNIT WEIGHT

pieces, either by passing under a knife whose blade is rotating at
a fixed speed, or manually by an operator. Each of these pieces
is nominally the same length, and each – since the cross-
sectional area is 'constant' and the material density 'unchanging' – is of the 'same' weight as its fellows. All these similar
pieces should weigh the same.

But they do not. Their weights vary, even from a process held
at 'constant', which is to say nobody is interfering with the
process controls – the pressure, the temperature, the timing
even at this 'constant' setting condition the weights of successive similar pieces are different.

THE PROBLEM

Each of these separate pieces will become, at the next production stage, a finished component. This is achieved by again
remelting the rubber so that it will flow under pressure and
pressing it into a die or mould. The volume of this die is designed
to produce the desired component, on the assumption that the
weight of the rubber pieces presented to it is correct, and

remains correct for every piece used. At this, *late*, processing stage, any extruded and cropped pieces which are *under*weight will be too light to fill the volume of the die and will produce an *un*filled component – a rejectable defective. Any pieces which are *over*weight will produce excessive 'flash' on the component, which has to be removed, and also will leave a residue of surplus rubber adjacent to the die; this rubber is from here on unusable and represents costly waste.

The problem facing the rubber converter was *excessive weight variation* beyond engineering tolerances, in the output of extruding pieces, resulting in both too many unfills and in too much wasted irrecoverable rubber at the die-pressing stage. This excessive variation in the unit weight of the cropped pieces from the extruder manifested itself within every delivery of two to three thousand pieces produced by the extruder as a 'batch'. This had been the norm for fifteen years. So the process was deemed 'delinquent'. It was the 'fault' of the raw material. It had 'always been this way'; and therefore, by assumption, would always remain this way? The MD believed that this need not be the case, he believed that with the right approach things could be changed for the better; that a new pair of eyes, naive to the technology and its traditions, might see through the smoke of the daily firefight and glimpse the solution that was there but was invisible to everybody else. He was right.

THE SOLUTION

Now that we are more familiar with the manufacturing process we can follow the procedure.

1   Select the *right* project. Tackle this one, and only this. For the time being. Aim to achieve *one* small victory. The project chosen was the manufacture of *tank-track links*. A tank-track is one of the pair of endless belts on which armoured fighting vehicles achieve mobility; it is made up of a series of hinged links, much in the manner of a bicycle chain. Each link is made of a metal casting with a hole in the middle (please see Figure 6.2); across each corner of this hole is a triangular web whose purpose is to retain the block of rubber injected into the hole. This forms the soft tread of the link

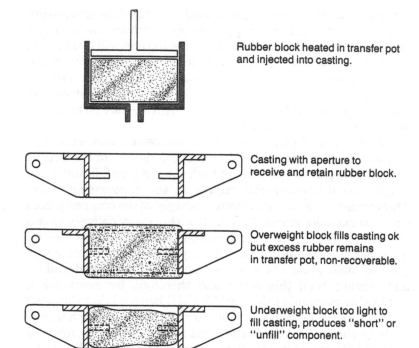

Rubber block heated in transfer pot and injected into casting.

Casting with aperture to receive and retain rubber block.

Overweight block fills casting ok but excess rubber remains in transfer pot, non-recoverable.

Underweight block too light to fill casting, produces "short" or "unfill" component.

**FIGURE 6.2   EFFECT OF INCORRECT UNIT WEIGHT OF
RUBBER BLOCKS**

which permits the vehicle to travel on metalled roads without ripping up the tarmac. The rubber block is inserted into this hole by positioning it under a 'transfer-pot' into which there has been loaded an extruded (through a square-section die), and cropped, rubber block of the 'correct' weight. It is then heated and squeezed into the cavity provided by the hole in the track-link. Sometimes the cropped blocks of rubber were the 'correct' weight, sometimes not. From within discrete production batches it was found that between 5 and 10 per cent of blocks were underweight, and causing 'unfilled' links. These were rejected (now that all value had been added you will note), and sent several miles away (which cost haulage money), to a company specializing in cryogenic deflashing, who immersed the links in freezing liquid nitrogen to embrittle the rubber, then banged it out with a mallet before fettling the casting (they were *paid money* for doing this) and returning it (which cost yet more haulage money) to the rubber converter to be done over again. A bottle-to-bottle hook up costing the converter large sums of money. Also an opportunity to show him how profitable good quality management can be.

Batches of blocks containing underweights were also known to contain an excessively high proportion of over-weights, and because these did not cause underfills their effect was less obvious and thought to be less costly; these overweights in fact proved to be the most profitable element of the solution to the problem, as we shall see. Remember – Quality is about making more *money* by making fewer mistakes, and by opening up more opportunities.

2   Give it *high* visibility. Turn this first project into a flagship job. Let it be known that this particular problem, a manager-ial bushfire which has been smouldering away and combust-ing the cash of resources into the ash of defectives for so long it's left a charred hole in the corner of the treasury coffers, is about to be solved. This has the effect of attracting an interested audience of sceptics, who've 'seen it all before', and will 'believe it when they see it', and who have already 'tried everything' and become convinced that 'there is nothing you can do about it'. These doubting Thomases are about to be shown that it can be done, and once they have tasted this little victory they will be thirsting for more. Do not blame them, be glad of them, their scepticism is well founded, but it can be flipped by success into an enthusiasm hot enough to generate the steam to drive the rest of the culture change which Total Quality Management is about to initiate.

3   Give it *high* credibility. Find out from Factory Accounting how much – in hard currency – the wasted do-over-again work of making wrong first time has been adding to costs. These are *visible* costs. To recover them by breaking the bottle-to-bottle hook up will recover not only these (directly to the bottom line), but also the other, *hidden*, costs asso-ciated with poor quality management which are submerged

in the swamps of the accounts. These total costs are usually at least three times the visible costs, so they afford potential for making *spectacular* savings and securing dramatic credibility. This all serves to make life that much easier later on in the game, when the going can get really tough. So, it makes sound tactical sense to stack up as many credits as possible very early in the game.

4   Form a *trinity* of alliance. The trinity, the three-in-one and one-in-three, the triad, can be a very powerful form of human association. This trinity includes the client company's quality champion (usually someone from the quality department). In the case of this company of rubber converters the Assistant Quality Manager, a young man fired with the need to make his mark, as are all young men, and astute enough to perceive potentially good career-mileage in this project, took on the mantle of champion. The second member of the triad is the consultant, who is there to act as mentor to the champion. The third member is a manager or director, somebody senior to the champion, a patron who is there to protect the champion's back from the sly stilettos of any political assassins who might be around. Remember, this is the real world: the quality champion is not the only person hereabouts with a spark of ambition; there are rivals around who for one reason or another might be quite happy to see the project fail whilst commiserating with the champion about its failure.

5   *Do it*. The consultant had already spent a few hours talking with (that means 'listening to' as well as 'talking to') the operators on the tank-track link line. He had chosen this line when the champion had asked 'Where do we begin?', to which he had counter-questioned 'Where does it hurt and by how much?' which is when this particular project presented itself to his trained eye like a pikestaff poking up in the cornfield of their conversation. He invited the operators to tell him, and show him, what they did to 'control' their process. This act of itself triggers off a 'Mayo reaction' or 'Hawethorne Effect' (as described in my book *Right First Time*). Here was a stranger, a consultant, pleased to listen to

each man talking about his favourite subject – himself. This open relationship generated high trust and a spirit of enthusiastic co-operation towards the common objective. *They* want things to be done more effectively. They too are fed up with producing junk, any junk at all, let alone anything between 5 and 10 per cent of it in every batch they produce. They are trying their best, by *all the means at their disposal*, to eliminate it. They are diligent and thorough in all that they do. They do not want to do a bad job, nobody likes doing less than a first-rate job.

So why were they getting it wrong? Using all the means *at their disposal*, and still not achieving excellence, what was wrong then?

Perhaps they had insufficient means at their disposal.

But they had a specification sheet, which informed them that these blocks shall weigh 800 grams, with a tolerance of plus and minus five grams. They had an electronic balance, calibrated to tenths of a gram, with which to weigh the blocks – OK, they were unable to weigh *all* the blocks, the output was too fast to permit that, besides they had other things to do. But the weighing machine was even wired into a little calculator which at the press of a key would churn out lots of numbers, the 'control' information was there, though some of the numbers didn't mean much to many of the people.

Why then was the weight variation of blocks, even from the same production batch, so excessive?

Was it *really* excessive? Or was the variation merely *normal*, 'a fact of life'? There is always one sure way of finding out: measure what is 'normal', assess the capability of the process in terms of weight variation. Any variation beyond this can then be validly described as 'excessive'.

Bring order. There will always be some variation in the weight of blocks of rubber, which are the outputs. This is because there is always variation to the inputs to the process. These include variations in polymer viscosity at 'constant' temperature and pressure, arising from the variations in molecular weight of polymer passing through the die at any given instant. There will be variations in the heat applied to the system as steam or electrical pressure varies within its natural limits. Likewise with hydraulic pressure. Likewise with *all* the inputs to the system, all are varying all the time. This variation, intrinsic to the process, is inescapable. It is a reflection of the laws of physics and chemistry. You cannot avoid it. The best you can do is measure it. Then monitor it.

Recall the Three Rules of Quality:

1   No inspection or measurement without recording.
2   No recording without analysis.
3   No analysis without action.

This is what the mentor/champion/operator team did, set the extruder machine to run and weighed a couple of dozen random blocks from the output. Then analysed the findings to determine the average weight, and the standard deviation of the weight which is a numerical measure of the weight variation of the process.

This process was found to be a very stable operation in terms of weight control, as the consultant had guessed. The *instability* which was giving rise to too high a proportion of both underweight and overweight blocks was *operator-induced*. The operators had been diligently using *all the means at their disposal* to control the process and try and make it produce blocks whose weight was within the prescribed tolerance band. But one very important tool was not at their disposal – statistical understanding of the nature of random intrinsic process valuation. They did not know. Nobody had told them. So whenever they weighed a single block and found it to be heavier than 800 grams, they diligently turned down the pressure so as to squeeze a little less material through the die between the knife-cuts, thereby producing lighter weight blocks. Next they would weigh one of the blocks they had caused to be made lighter, discover it to be underweight and in alarm reverse the procedure and turn pressure up to make them heavier. They were pursuing the natural variation of the process, and rendering it more variable by their intervention. This is called *hunting* a process variable. It is a very prevalent fault, which is costing industry huge and unnecessary costs, perpetrated by the well meaning who just happen to be inadequately educated in statistical thinking. This operator diligence was creating all the overweights and underweights; it was taking the natural rhythmic variation of the process and multiplying it threefold. This was the introduction of a 'special' or 'assignable' cause to process variation, in this case 'operator over-tuning'.

The answer was so simple, so obvious and so effective. Give the operators a control chart, with limit lines based on measured process capability, which not only fulfilled the Third Rule of Quality by telling them *when* to act, but, more importantly told them, when *not* to act. 'No action' is an action decision too.

This was done. The results, measurable in the pounds sterling saved by making zero unfills, were gratifying. The 'delinquent' process had become compliant. The 'faulty' raw material was seen to be all right. The 'Black Art' was unmasked and shown to be a figment of managerial imagination. This was good; there was better to come.

Quality is not 'book learning'; it is essentially the most *practical* of crafts.

'In the past, when you encountered unfills, what did you do to try and get out of the problem?' enquired the consultant. 'Did you jack up the nominal target weight so as to make fewer underweights?'

The client affirmed that at one time, when faced with over 20 per cent of unfilled links, they had indeed upped the target weight, which was how they had reduced the incidence to between 5 and 10 per cent.

He advised them to reduce the target weight, in controlled 10-gram steps, until they met the threshold at which underfills appeared.

They reduced it, even though it went against the grain of the 'custom and practice' cop-out of their conventional wisdom, and managed to reduce it from 800 to 765 grams. The savings in rubber hitherto gone beyond recall more than doubled the money savings already made. Straight to the bottom line.

Then there were the hidden costs in that whirlpool of waste . . . we shall talk of the whirlpool – the jackass's jaccuzi – later; since this is where the money goes, this is where the money can be recovered.

So this is one way of implementing Deming's Fifth Point, 'to improve constantly and forever every activity . . .' but not by trying to do all things at one fell swoop, not by trying to be all things to all men at all times. This phased attack, to implement the total system on a project-on-project, wave upon wave approach, building bigger successes onto small successes, is a tried and proven way of doing it. Remember, too much too soon can be as bad as too little too late. The timing of the initiative and its pacing are vital to its success.

The Idiot's Tale: 6
Bedlam were launching a new *greenfield* operation! Big Boss was elated but apprehensive, this was a pioneering technology, smelling of unforeseeable pitfalls, and his reputation – and, more important, his self-esteem – hinged on its success. Success meant running the operation up on a schedule to supply masses of components to a single contracted customer organization renowned around the world for the rigorous demands of its quality. 'I want you to appreciate that these people are professional bastards,' he was generous enough to advise me the day before a squad of the hit men arrived. They were hard characters, these. They asked awkward questions . . . 'What will you do when we send you your first million components back?' they enquired with a silky menace. Big Boss asked them why they would commit such a hostile act of rejection. 'Quality', they said. Big Boss pointed towards me. 'He's the quality man,' he said, and the gun-muzzles swung onto me. 'We shall more than meet your quality standards,' I assured them; 'either that or we shall not send you any deliveries. If we discover that our process is not capable of meeting your specified requirements, we shall negotiate.'

They were having none of this kind of backchat. 'Pompous bullshit,' they sneered, 'it *always* happens. And what will you do with the *second* million we reject? *That* always happens as well.'

Big Boss wasn't enjoying the tone of this conversation, but he too was tough, and loyal to his men. 'If *my* quality man tells you that you'll receive nothing but output of unimpeachable quality then that is *exactly* what you will get.' Big Boss had this peculiar ability to *expand* when he got angry, he actually appeared to inflate like the Incredible Hulk. He was angry now, growing bigger like an expanding universe of indignation, swelling fit to fill the boardroom. Good old Big Boss, nobody messes *you* about, but please don't burst.

They persisted in their interrogation. 'Let's go out and have a drink,' they said, 'talk things over.' Big Boss gave me the nod and the wink which meant 'Don't worry about spending a few pounds' and I, the quality man, hosted the two hardest of the hit team in the local watering-hole used by the big game.

'We drink whisky,' they informed me. 'Doubles,' said one. 'Trebles,' said his accomplice. 'So do *you*,' they ordered. We drank whisky, which has never been any problem to an enigma such as me, so we drank some more. When they were good and ready the knives came out. 'We have *destroyed* incompetent quality men like you, did you know?' they attacked. 'Now answer the question – what will you do when we send you back your *first* million, and your *second*, and your *third* . . . What will you do?'

'Fight you, you bastards, what the hell else do you think I would do?'

They fell about laughing, nudging each other in the ribs, telling each other, 'He's passed the test, he's OK, we shall get on all right with this one.'

'You see,' they explained, and I could comprehend because anger keeps you sober – adrenalin being stronger than alcohol – 'when we reject back onto a supplier we *don't* want meek compliance, yes sir, yes sir, three bags full sir; we want *resistance*. *You* are the grinding wheel on which *we* keep ourselves sharp. Agreeable wimps are of no use to us, we just had to make sure that you weren't one. Have a drink, our round.'

This encounter with the selection committee of the hard men marked the beginning of a joyous and fruitful relationship. These men were hunters, Athenians, lean and sharp. They refused to do business with lesser mortals.

There must be a moral somewhere in this. It certainly resulted in zero-rejection shipments, high productivity, and ongoing quality improvement to both process and product. Absolutely concordant with the things Deming says.

Summary

The Welsh language is rich in words of wisdom. There is an old proverb – 'Nid da lle gellir gwell' – which translates into 'It is not *good* if it could be *better*'. Perhaps Dr Deming has heard of it, and made it into his fifth point – 'improve constantly'. It exactly expresses the same sentiment.   ˙

How this approach was used to stunning effect in the operations of a rubber-processing company. This finds expression in the project-by-project system of gradual and unrelenting quality improvement. Start simple. Start small. The journey of a thousand miles begins with one step. Form the trinity of the tri-partite alliance, and use it to bring about improvements, continuously. Observing the Three Rules of Quality – measuring, recording, analysing – and then acting.

Quality at Bedlam's new operation and the demands of the hard men, leading to the foundation of a relationship of total integrity and complete mutuality of benefit.

7 DEMING'S SIXTH AND THIRTEENTH POINTS

6 Institute training and education on the job, including management.

13 Institute a vigorous program of education and re-training. New skills are required for changes in techniques, materials and service.

These amount to much the same thing in the end, so we shall examine them jointly in this chapter.

Training and education for management. Why? Because they are a Good Thing? Because sitting in a classroom playing management games with monopoly money beats working as a way of pleasantly passing time? Because being singled out and sent away to some business seminary sets you apart from the hoi polloi? Because being let out for a while to ride on wisdom's wings for a week or two is a kind of legitimized truancy? Because yours is just another bum on just another seat, like everybody else's; and they are all there, fidgety with the pins and needles of unaccustomed inactivity, for one purpose only – to help the Management Development Manager achieve his target annual quota? College fodder? Is *this* why? Sometimes it is. Or is it to help you do a better job in the humdrum grind of your daily work? It's supposed to be.

Have you ever been on a 'Management Development Programme'? Can you recall the experience?

It's round about teatime on the first Sunday of the programme; you arrived a couple of hours ago and were shown to the spartan cubicle which will be your nocturnal hidey-hole for the next twenty nights (this is a *young* managers' course). At the moment you and fifteen others like you are standing around in your concertina-legged trousers with your jacket buttoned up, your left hand out of sight behind your back after the manner of a Queen's consort, your right arm sticking out rigidly horizontal from the hip to save spilling the contents of the half-pint pot which you are cradling as cautiously as if it were the Holy Grail. This is the time of the pre-dinner drinks to the pre-course dinner. Calculating eyes flitting with uncertain inquisitiveness, you are as wary as a pack of stiff-legged dogs eyeing each other up, sniffing the signals, sensing the subtleties of status, exploring the ranking order. Where do I fit?

You find your place at the Big Table ready for the Big Meal, an army of spit and polished silverware drawn up into platoons on a parade ground of snowy linen. This is their shock-troop induction into that over-abundance of victuals which is college-fodder. You are learning how to behave in a situation of social stress; this food is not to nourish you, it's the props of a semi-religious ritual.

Later that evening, after the Big Man who is your ultimate corporate boss has uttered his uplifting homily after the Big Meal, you discover the Billiard Room. Now *that's* more like it. This is familiar territory, and green baize is the same colour the whole world over. Soon the snooker balls are banging, friendly games are going. All those misspent years of your spotty adolescence, stooped over the green cloth squinting down the ivory shaft of your 18-ounce cue with its green-chalked tip, were not wasted after all. You used to be good at snooker. You spent so much time playing it that you even began to develop the pallid tinge of the professional, that consumptive look which comes from labouring too long in too little daylight. During the next three weeks you spend every moment that can be spared from lectures and playing the computer-simulated business game at the snooker table. You relive the life of freedom you enjoyed before you got married, you reoccupy that universe of stratified blue cigarette smoke and spinning, coloured orbs, a universe whose silence is broken only by the continuous clicks of balls colliding with other balls, the thumps of balls bouncing off cushions, the clinks of balls dropping into pockets to land on balls already in the bag.

On your last night your boss arrives, to attend that funeral wake known as the End of Programme Dinner. 'You've put on weight,' he remarks, and you tell him about the food. 'What did you think of the course then?' he asks.

'It was just a load of balls,' you say.

You *must* remember, it's not all *that* long ago.

Things have changed since then, though – well, they have changed in some places; perhaps there are still pockets of reaction here and there where education is being offered in the wrong subject to the wrong people for all the wrong reasons, who knows?

It is interesting to note that Deming's points mention both 'training' and 'education'; clearly he sees these as being two differentiated activities. Let us try a couple of definitions . . .

Training is the imparting, by systematic instruction, of a set of skills. Its purpose is to equip the trainee with the ability to *do* whatever it is the job requires to be done if it is to be performed effectively. It is specific, task-centred, and tightly focused. It is therefore minimalist (how little needs to be learned? It has a *need* to know basis), it is measurable and it is finite, close-ended. A person is *subjected* to training, is a *recipient* of it. Training is limiting and confining, it is bounded.

Education is the sharing, by organized disclosure and dissemination, of knowledge, understanding and insight. Its purpose is to open and widen the mind of the pupil, to search out and explore the general principles which underpin the observed phenomena. Its purpose is to enable the person to become an *enabler*, one who can enable others *to do* (those who have been *trained* to do). It is generalist, non-specific, of diffuse focus. It is maximalist (how *much* can be learned?) It has a *good* to know basis. It is assumed to be measureable. It is infinite, open-ended, unbounded. A person is *exposed* to education (or rather to education-promoting influences), is an *experiencer* of it. Education is unshackling and liberating.

Whether you agree with these two definitions or whether you want to pick the nits out of them is neither here nor there at this stage, because defining things like these is akin to embarking on those words-without-end debates enjoyed by the drinktelligentsia during the pontification stage of getting drunk. As far as we are concerned these definitions are valid enough to serve our present purpose, which is to discuss their relevance to the teaching of Total Quality Management in pursuit of higher productivity and quality. It is important that we differentiate between the two and keep them well apart in our minds even though they sometimes serve as the two strands in the one rope. It's when the two become tangled in the mind, and one is confused with the other so that we forget which is which, that the trouble starts. All education involves some training, but not all training requires education.

So what is training *for*, and what is education *for*?

Training is the discipline imposed by the technology upon the practitioner. Training is the acquisition of a repetitive skill in a process of fragmented work. It is firmly placed in the d-realms of work (Deficiency, working simply to earn enough to meet the needs of existing).

So is it possible to *train* a manager? Only, one suggests, in the technology of the enterprise he is managing. Is it sensible to speak of *training* a manager in the *un*scientific art of management? Or do you believe that 'management' is a branch of science, a kind of 'people-engineering', that there are 'laws' of management as there are 'laws' of physics, astronomy, valency, gravity . . . ?

Education is the gateway to liberation. So you had better be careful whom you educate, and what you educate them about.

Education is also an act of supreme faith and trust on the part of the educator. 'What will they do with this education?' the teacher wonders, and will rarely find out.

People, Things and Systems – The Three Props of Industry

What is the *nature* of the training and education and retraining Deming speaks of in his sixth and thirteenth points? To give us some sort of structure to hang our enquiry on let us consider the three-legged stool (see Figure 7.1).

The left leg of the stool is called 'technology', it is to do with 'things'; the middle leg is marked 'management' and is about 'systems'; the right leg is labelled 'leadership', and this is concerned with 'people'. These three legs are joined in common purpose – to support the seat of the stool on which rests the

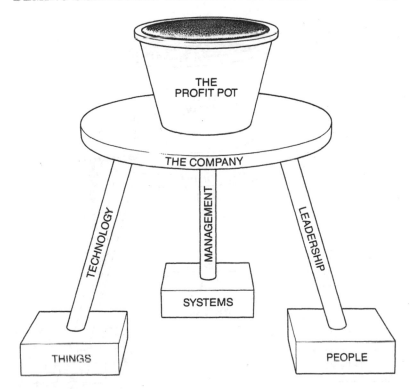

FIGURE 7.1   A BALANCED COMPANY

enterprise. If the enterprise is to sit securely then it is as well if the three legs are all more or less the same length. If any of these three legs is shorter than the other two the stool tilts and the enterprise finds its perch uncomfortable – it tends to slide towards trouble where the leg is shortest. Though a 'stability' still exists the whole organization is canted out of balance. Balance is important – there is a basin on top of the stool (see Figure 7.2).

This basin, being basin-shaped, is wider at the top than at the bottom. It is able to contain a certain maximum volume – rimful – of a certain liquid; a rich and golden nourishing liquid called 'return on investment (ROI)'. It is impossible to fill the basin beyond its designed capacity, but *it is only too easy to cause it to be underfilled*. If the stool tilts out of balance, the basin tilts with it; this, because the level of a liquid preserves horizonta-

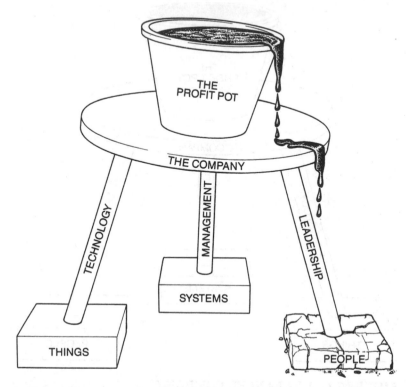

**FIGURE 7.2   A BADLY-LED, UNBALANCED, COMPANY DRIB-
BLES ITS PROFITS AWAY**

lity, reduces the capacity of the basin. The volume of liquid it is
potentially capable of holding cannot any longer be contained
and some spills over the lower side. Even the slightest tilt is
enough to reduce the basin's volume by an amount dispropor-
tionately larger than the angle of imbalance; because the basin
is widest at its rim, a lot of liquid dribbles over the edge and is
lost in the direction of the shortest (weakest) leg. Luckily there
is still a lot of liquid remaining in the slightly tilted vessel, but
the cream is skimmed off. So it is quite possible for an organization to be a few degrees out of balance, and therefore making a
lesser ROI than its maximum potential capability, yet still
regard itself with justification as 'successful'. (The work of the
spirit level, measuring balance, is the job of the MD.)

Until the shortest leg gets shorter to the point where the angle of tilt overcomes the friction between the basin and the stool, which precipitates the slide into bankruptcy.

The aim of running the business is the attaining and the maintaining of this balance. It is not a *static* balance like, say, a tripod designed to hold a theodolite steady on uneven ground. It is more a condition of *dynamic* equilibrium like, say, a juggler balancing a dinner plate on a cane. Which means it demands constant attention if it is not to topple.

This is where training and education come in. To redress the balance.

Now to have named these three legs Technology (things), Management (systems) and Leadership (people) is quite arbitrary, and it has only been done for the sake of convenience in communication. This is just another 'concept', that is all; and a concept is in no way the statement of an absolute, it is nothing more than a manmade vantage point from which we might glimpse a facet of understanding. That is what concepts are *for*. To call one leg 'Technology' is not to exclude 'people': people *devise* technology. To call another 'Management' as if it leaves out 'people' is unreal: people *invent* systems of management. But this concept will serve us well in our considerations of training and education. Especially so if we tie it in with yet another – that of the 'honest serving men' of Rudyard Kipling, who wrote (in the *Just-So Stories*),

> I keep six honest serving-men
> (They taught me all I knew);
> Their names are What and Why and When
> And How and Where and Who.

What, why, when, how, where, who; perhaps it might pay us to press these hearties into our service, though not necessarily in this particular sequence.

Applying Deming's Sixth and Thirteenth Points

WHERE DO WE INSTITUTE TRAINING AND EDUCATION?

In whichever of the three directions the stool is tilting and the basin spilling. Is the outfit weakest on the things, system or people leg? Obvious, isn't it? Simple. Sometimes too obvious and too simple to some people.

In the real world of doing things for money, either manufacturing articles for sale or selling a service, it is very often the case that nothing is quite what it seems to be. Nothing is really simple and the obvious is often as insubstantial as a mirage. Consider the nature of the obvious; it must be very obvious to you by now that what is obvious to you is not obvious to me. 'But it's *obvious*, isn't it?' we ask our colleague who is acting dumb, our voice rising to match our mounting frustration at his obdurate refusal to see the 'obvious' which is staring us in the face as clearly as our reflected image stares at us from the looking glass. Our colleague is getting just as frustrated as we are, 'Surely *this* is obvious . . .' and each of us is seeing our own 'obvious' because *each of us is seeing what we believe*. A bit like when you look into the mirror. Not 'believing what he sees', that also would be too obvious and too simple, so it's the other way round, we see what we believe. So which of the three legs is letting the enterprise down? Perhaps it is no longer as 'obvious' as it seemed at first sight.

(Later on, in the Idiot's Tale to this chapter we shall use again the example mentioned in the Idiot's Tale at the end of Chapter 4 – the plastic food pot saga from Bedlam. This will illustrate these points quite nicely.)

HOW DO WE INSTITUTE TRAINING AND EDUCATION?

That depends upon whether we are teaching *skill* (training), or *knowledge* (education). Are these two – skill and knowledge – the same thing, or are they completely different? To answer this let us recall some of the skill and knowledge associated with the practice of medicine in its early years, say around the time of the Battle of Trafalgar, in the first decades of the nineteenth century.

Then as now, skill was a tradeable commodity and practitioners aspired to achieve 'state of the art' skills. The principal skills in their sort of surgery lay in the *speed* of execution. The most eminent surgeons were those with the sinewy arms of a blacksmith and the manual dexterity of a cardsharp. These qualifications enabled them to amputate a gangrenous foot by making a couple of quick slashes through the calf-muscles to the bone beneath, with flashing carving knife; rasping through the shinbones in a few urgent strokes of the shipwright's saw; secure the flap of skin over the severed end with three or four fast stabs of the sailmaker's bodkin; and seal the wound with a quick hermetic dip into boiling pitch. All done in less than twenty seconds against the stopwatch. Some of the patients actually survived. Without this state-of-the-art treatment all would have died. That was *skill*. Was it *knowledge*? It embodied the knowledge available at the time, but the technology of surgery (technology equals scientific knowledge applied to the affairs of business; a fair definition?) was still awaiting the advances to be made available to it by the body snatchers and their dissectionist clientèle. But even without this knowledge the surgeons of the time exercised considerable *skill*. So skill and knowledge are not the same. Skill is the product of practice, and takes time to acquire. Knowledge is immediate, one minute you do not know, the next minute you know. If it is to be effective, skill calls for coaching; knowledge needs nothing more than the imagination and the opportunity to express itself.

Sometimes we still tend to confuse the two. We send people away on a course to learn something. Forgetting that the purpose of the course is to impart a *skill*, we are disappointed to discover that the pupil is able to remember little more than about 20 per cent of what was taught. This is why Deming mentions 'on the job' training. Skills should be taught on the job, and then always reinforced with coaching. Knowledge can be taught anywhere. If it is interesting it will be remembered.

WHAT DO WE TRAIN AND EDUCATE ABOUT?

We teach *statistical thinking*. We are said to be living in the Information Age. It could be called with greater accuracy the Age of overabundant Data. We are suffering from data-overload. Overload induces stress, stress causes mistakes. A lot of the 'information' is in the form of numbers. Numbers have been known to intoxicate; number-drunkenness is yet another of the addictions which imperil performance. Numbers, extracted as *data*, from our business processes (*all* our business transactions, not only in mass manufacturing), do not mean much until they have been analysed. Data is no more 'information' than a field of barley is a bottle of malt whisky. This process of analysis is effectively a means of distillation; raw numbers are reduced and distilled into their essence – Information. The only discipline to handle large numbers of numbers and make sense of them is *statistics*. That is what it is for.

Statistics as applied to business is a blend of skills (a trained facility to handle numbers economically, to get the most information out of the smallest amount of data); and of knowledge (an educated judgement able to distinguish between which data is relevent and which is not, and able to determine how much of it is needed to suit the purpose in mind). There is no other branch of learning equipped to take on this work.

So we teach *statistical thinking*. We do not teach *mathematical statistics*. We do not want a workforce of statisticians. We do want a workforce educated in statistical thinking, trained in a few powerful statistical techniques, confident and competent in applying them.

The aim of the business is not to produce statisticians or statistics, it is to provide a reasonable return on investment (a return expressed not only in terms of money, but in many other terms as well).

So statistical thinking, statistical awareness, is the *content* of what is taught. But we have been teaching this for fifty or more years, to little avail. We need something to go with it if it is to find effective expression. For decades we have been concentrating our training on the technology leg and the systems leg, focusing our attention on the inanimate, and neglecting the third leg. So we have toppled into imbalance, and the remedy is to teach things to do with this leadership–people leg, to impart awareness of the *context* within which the skills of statistical techniques will be used to best effect.

WHEN DO WE TEACH THESE THINGS?

All the time, ongoing – to embark upon this kind of training and education programme is to begin a journey which has no end. It becomes a way of life. The training purist will argue that we should first identify a training need, then prepare a training programme to meet that measured need. Fine, OK, we need *systematized* training such as this, if only to be discriminating in what we teach. But systems can become too dearly beloved of those of us who no sooner devise a system than we turn it into a totem pole. The system itself will become the 'thing', a discrete entity with an existence of its own, its teaching purpose eclipsed. Formalized training is good, but the informal training which you can use to follow it up is better. It becomes as quick and powerful as gossip. It goes like this;

Teach something to two people. Drum it in. Hammer it home. Now tell them each to take two other people and tell them what has been learned. Then tell them to ask their pupils to do the same. It sets in train a process of doubling:

One teaches two, two teach four, four teach eight . . . in nine steps the word has been spread to five hundred listeners, in ten steps to a thousand. That's how gossip gets around the organization so fast. Unless it is kept clear and straightforward the original message gets a bit diluted and maybe discoloured as it goes through the process, but you can make good use of it even so.

WHO GETS TO KNOW?

The answer has to be anybody who in any way at all is able to affect the performance of the company. Which is everybody. Can you think of anybody in your company who is *powerless* to affect its performance one way or the other? If you can, then why are they still on the payroll? Total Quality Management excludes nobody. From the operator on the switchboard who takes the call that brings in the order, through the entire processing network, to the lorry driver who delivers the goods or to the employee who provides the service, *all* have a part to play and all will benefit from the Deming Doctrine. The whole of business is a flow process, a stream of information; so it is amenable to being improved by the application of SPC – statistical *process* control. (Not Statistical *Product* Control; control of the quality of the product is achieved upstream by controlling the process to its own best capability, *any* process.) SPC is one of the tools of TQM, which is essentially a philosophy of service and a discipline of frugality. Which is to say, a creed of good management.

WHY SHOULD COMPANIES IMPLEMENT DEMING'S SIXTH AND THIRTEENTH POINTS?

Simple, because any company's competitive edge is rooted very firmly in the leadership/people leg of the stool which supports the enterprise. Admittedly it is possible to secure a competitive edge through the technology, which is why Research and Development are such important activities and the striving for innovation is so important. But the findings of R & D and the exploiting of innovation are very much in the leadership leg of the business, the people bit, because of scaling up from laboratory to plant operation. But even before this, during the process of research itself, data has to be turned into meaningful information if it is to point the way to the next – the development – phase. The only tools for unravelling tangled data are those provided by statistics. Even after the successful completion of the development phase into full-scale production, the process can be kept on track only by the use of statistical methods.

So in answer to Why? Because there is no other way.

The Idiot's Tale: 7
Let us talk some more about Bedlam, and *their* plastic food pot with print problems, which I told you about at the end of Chapter 5. Too many of the pots *they* printed had blotches on the decorated surface. Was this a problem of Technology (things), of Management (systems), or of Leadership (people)? In other words, which of the stool's three legs was shortest and so causing the basin to tilt to such a degree that all ROI was cascading into the sands?

The Technology leg? Defective print is a problem to do with the technology of applying ink to a thermoplastic surface, isn't it?

But the print-misses were caused by the presence of smears of oil, contaminating the pots at the injection – moulding stage. 'This is a technological problem,' confirmed the Production Manager in charge of moulding the pots, 'not my problem, theirs, the Technical Department's.' Well, he would say that, wouldn't he? He added, 'Anyway, they all get printed, so the printers can pull the duds out, can't they?' The self-excusing cry of the upstream polluter. 'I'll pour the poison in and it's up to you to filter it out downstream.' What, at a speed of four thousand units an hour?

OK, a 'technological' problem, is it? Then let us ask the Production Manager a question. 'Where does the oil come from that gets onto the pots?' The PM, a man celebrated in the factory folklore for his sour wit, replies, 'The moulding machine has moving parts. They have to run smoothly otherwise they will seize up, so we use what we call lubricant to make 'em run smooth. You would call it oil.' As he says, facetiously, it is a technological problem. So far . . .

'From where does the oil get onto the pots?'

'From the tie-bars, as the pots are being ejected – they catch against the tie-bars and sometimes there is oil on the tie-bars so it gets onto the pots.' He knows his stuff, this Production Manager.

'The tie-bars are not moving parts, so they don't need oil; why is there sometimes oil on the tie-bars?' The problem seems to be no longer technological.

'It drips onto them from oil seals higher up the equipment.' He has an answer for everything.

'Why does it drip? Are the seals faulty?'

'A little bit, they spring a very slight leak now and again, they cannot be made totally oil-tight. That's how oil-seals are.' So the problem is back in the technology sector.

'Then wipe it off the tie-bars. That way it cannot contaminate the pots.'

'Can't do that, you have to stop the machine to wipe it, that means stop producing, that means lower production figures, that means it cannot be done.' The problem was no longer to do with Technology, it was now to do with Management (systems).

'So it's OK to make output which gives rise to downstream junk?'

'None of my business. I am the Production Manager, Mouldings. My job is to optimize the *production* of mouldings, to make as many as I can, it says so in my job description. That is what I do.' The problem was still in the management leg of the stool.

Time to move it.

'That is what you do, eh?' The time had come to exercise some hard power (we shall be talking about the nature of power later on), none of your sweet reasonableness any more. 'Not any more, you don't. From right now we impose Quality Assurance sampling checks on your output. Any trace of oil on the pots making it unfit to go to your customer, the Printing Manager, and your machines will be stopped.' The problem had now been shifted into the Leadership (people) leg of the stool, now it could be tackled.

Stop the machines! This is sedition. Reduce the output figures! This is sabotage.

There was a simpler solution, which was to ask the

Production Manager, 'Why don't *you* just stop your
machines, routinely, every two hours or so, and wipe away
all traces of oil from around the seals before it has had time
to accumulate and seep downwards until it starts to drip
onto the tie-bars? That way it will never be in a position to
pollute the output of pots. It will only take a minute or so.'
  There was more to it than this, but this is the essence of
it. Commonsense and Communication.

Summary

Quality can never be, and must never be looked upon as, a magic
dust which, sprinkled on the process, will resolve all problems.
There is no magic available, though the outcomes of doing
things the Deming way can sometimes seem near-miraculous.
But if there is no magic ingredient in the recipe for success there
is certainly a key element without which it will surely fail –
training and education.

  This is the thing Japan has used down the last few decades to
detonate its economic explosion – education. Quality circles are,
in fact, principally devices for bringing education to workers
traditionally denied access to education by the precepts of
Taylorized 'scientific management'. This is the 'people' leg of
the three-legged stool which is a metaphor of the organization.
Educating in statistical methods to enable management (and
the workforce) to distill meaningful information from process
and product data. And then use it to control the process.

  How this concept was applied to a problem at Bedlam plc and
yielded complete success.

8 DEMING'S SEVENTH, TENTH AND ELEVENTH POINTS

7 Institute supervision. The aim of supervision should be to help people and machines to do a better job.

10 Eliminate slogans, exhortations, and targets for the work force asking for zero defects and new levels of productivity. Such exhortations only create adversarial relationships; the bulk of the causes of low quality and low productivity belong to the system and thus lie beyond the power of the work force.

11 Eliminate work standards that prescribe numerical quotas for the day. Substitute aids and helpful supervision, using the methods to be described.

Some of Dr Deming's disciples say that occasionally his words express 'lofty philosophical concepts', implying that these might be beyond the intellectual grasp of lesser minds. Or maybe they don't understand what he is getting at so they dignify the communication gap by calling it 'lofty'; on the other hand perhaps they hear quite well what he is saying and feel inclined to disagree with it, but fail to do so because disciples don't openly sow doubt about holy writ. There is no room for Doubting Thomas in their dogma. Theirs, they say, is the one and only road to salvation, all other paths lead to perdition.

This, if it is the case, is a sad state of affairs; sad, but all too human, this degeneration of a healthy doctrine into a sickly dogma done by the faithful. This phenomenon has a very important bearing on the affairs of quality, and we shall be looking into it later on. In the meantime back to the points to be covered in this chapter.

Supervision

Let us have no misunderstanding of the meaning of this word, and no dilution of its purpose. One thing is for sure: though Deming sometimes may speak – according to his followers – in 'lofty philosophical concepts', this cannot be one of them. Lofty? The entire notion of overseeing or supervising is a base and demeaning concept; this is why he immediately goes on; 'The aim of supervision should be to help people and machines to do a better job.' He is saying that the job of supervision is no longer supervision; it has become something else, it has discarded the whip. It has changed.

Into what? What is its new name and role in our business organizations? During the first phase of its metamorphosis it turned into 'front line manager'. 'Front line' because it all happens at the sharp end on the concrete where all value is added, at the front line of the daily battle; and 'manager' because it sounds better than supervisor. Besides, management is different to supervision, managers manage and supervisors supervise, don't they? No, they don't. They should, but in practice very often they don't. Managers do not manage (generally speaking), they supervise supervisors, who supervise the workers. In their turn they are supervised by higher managers, who also supervise the supervisors and the workforce. They in their turn are supervised by directors who also supervise . . . big fleas, little fleas and lesser fleas . . . *all supervising*. Because they act, under the mindset of Old Order management, as if management is nothing more than glorified supervision, *overseeing*.

The term 'supervisor', as a description of what the job itself entails, these days is obsolete. So is the word 'manager'.

Are *you* managed? In your daily work do you enjoy being *managed*? Wouldn't you rather feel that you were being *led*? Wouldn't you rather *lead* than *manage*?

Leadership is better than management. It is so much more

satisfying to both the leaders and the followers.

The most modern term for supervisor in the organizational vocabulary is 'front line leader'.

What's in a name? you might ask. Plenty: names can exert profound influence on perception, behaviour and fortune. Consider the case, reported years ago, of an office clerk called Pratt. Now there is nothing wrong with the surname Pratt, the North West Wales Telephone Directory lists thirty-one of them, worthy people including an accountant, a fish merchant, and a vicar, who go about their lawful occasions untroubled by any connotations which might attach themselves to their perfectly respectable surname. The office clerk felt differently. He got this nagging idea that his name was somehow holding him back, so he changed it and had himself called 'de Havilland' instead. You can hear him answering the phone 'de Havilland here', all brisk and businesslike. Impressive, you must admit. If a voice on the other end of the line tells you it's de Havilland who is speaking you automatically pay it a bit of respect. Make aeroplanes, don't they? He certainly made progress, our Pratt cum de Havilland. Soon he was to be seen sporting a snappy pinstripe suit, and got himself promoted; at this point history loses sight of him. And just to show there is no discrimination in this discussion about the merits of certain surnames there are no less than three hundred and fifty entries under the surname 'Price' in the same phone book; it is easy to remain a mediocrity when you are part of that mob. What's in a name?

The English upper classes have long known the importance which attaches to names, which is why their daughters are nearly always given Christian names ending in 'a' – Georgina, Arabella, Claudia, Lucinda – you don't hear names like these being hollored across the treeless wastes of the council housing estates. They endow their sons with even more enchanting names, such as Ranulph-Twistleton-Wickham-Ffienes. With a name like that what *could* the boy become but a famous explorer? He is the man who circumnavigated the earth *north-to-south*, pole over pole. He has some interesting words to say about *leadership*, and he must be one of the best qualified men in the world to speak on the subject. He relates how, during his service in the Arabian deserts while he was a member of the SAS Regiment, he had under his command soldiers of three

nationalities who understood only their own language. Communication could not have been easy, yet the fact that he led his
polyglot warriors to success on several engagements of the sort
the SAS exists to tackle is proof of the effectiveness of his
leadership. 'Leadership,' he states, 'is about communication',
and he should surely know.

So the people who used to be supervisors are now front line
leaders. If they are leaders they are communicators. Over the
past few years many surveys, asking managers to rank in order
of importance several factors which influence their effectiveness, have shown that 'communication' is held to be the most
important. This is interesting; a further survey carried out to
find out how well managers communicate revealed that while
managers stress the importance of communication they hardly
ever say anything to anybody. Oh, they *talk* . . .

So what do managers do? They attend meetings. The luckier
managers are empowered to *call* meetings. The luckiest are
allowed time off to go to conferences. Meetings are for talking,
so during meetings managers talk but without actually saying
much. That way you can keep on holding meetings. They
sometimes form meetings involuntarily, in some kind of autono-
mic response to each other's proximity; you can observe it
happening in factories. Like an astronomer witnessing the
formation of a stellar cluster you can see something similar
taking place on the shopfloor . . . A solitary manager doing his
MBWA (management by walk about) pauses for a moment.
Another, lesser manager, many paces distant, senses the gravi-
tational pull and is involuntarily impelled towards the first
manager; now they form a binary system. This exerts stronger
attraction on any managers moving in more distant orbits, and
they too are pulled towards the original pair to form a planetary
cluster. The cluster enjoys a brief life before breaking up ready
to reform the next time the planets are in proper conjunction.
What has all this to do with leadership and communication?
What are they talking about? It's anybody's guess.

FRONT LINE LEADERSHIP

Can people be *trained* for leadership, or must they be *educated*?
Is it a *skills*-based or a *knowledge*-based activity?

It is both.

Are leaders *born*? Or are they *made*?

Without entering into that endless nature *v* nurture debate one thing is for sure, a person may be born with certain qualities which might grow into that particular ability we call leadership, but they will not grow into it without training and education along the way. The 'training' might be by just doing it, acquiring the skill by repeated practice as you go along. This, we are told, is how Moses learned his. His first attempt at it, when he punched an Egyptian overseer for being too liberal with the lash, was met by a rebuke from the chastised slave whose interests he had been addressing. Just who did he think he was anyway, was the indignant response from his fellow Israelite. No thanks for the youthful Moses. This is one of the essences of leadership; the true leader is not looking for thanks, he is looking for followers. Not because he 'wants' to be a leader – just about the best reason in the world for denying anybody access to any position of leadership is his expressed desire for it – but because he cannot help it, he simply has to do it. He *needs* followers. A leader without followers is a nonsense. Leadership is a transaction entered into and maintained voluntarily. It cannot be enforced, only authority is enforceable, and then not for long. It is based on mutual trust, respect, and *communication*. It can be learned, therefore it can be taught. It can be taught at the skills level and at the knowledge level.

If 'leadership' can be taught, can anybody learn it, and having learned it can anybody exercise it? Everyday experience and observation tell us that the answer to this question is 'no'. But perhaps everyday experience and observation have misled us; in looking around us for evidence of *leadership* we have instead seen only *leaders*. The most effective leadership is often a low-profile, near-invisible input ('Of the best leaders, men say "we did it all ourselves" ' goes the old Chinese saying), whereas we expect leaders to be highly visible. We are a bit like people who, beause they cannot *see* the wind, would contemplate a fallen oak blown down by a storm, and ask why the tree was lying on the ground. We tend to associate leadership with highly visible elitism, with 'charisma', and since charisma means 'a grace vouchsafed by the gods upon the chosen' we assume it to be a rare and precious thing, a kind of glittering cloak like a toreador's sequinned 'suit-of-lights', to be worn only by the elect few. We even allow ourselves to be dazzled by the very idea of it. Yet leadership is a commonplace activity.

If it *is* commonplace, then what *is* leadership? Let us try and define it.

Leadership is the imposition of one person's will upon the actions of others. Just as . . .

Power is the imposition of one person's will upon the actions of others. So . . .

Leadership is Power.

All leadership is power, but not all power is leadership.

And as the SAS explorer said, Leadership is Communication. Therefore Communication is Power.

Leadership is a social transaction, dependent upon communication, between the leader and the followers. It cannot exist in a social vacuum, it is exercised in *situations*. The situation might be a small group of people working together on a production line; decisions have to be made, actions have to be ordered, co-operative endeavour maintained, discord minimized . . . these are the jobs of the 'supervisor' who is nowadays known to be a front line leader. This job calls for skills, and *skills can be learned by training*. Skills are acquired through practice; learning is by doing.

This kind of skills learning is accomplished by practising leadership skills in contrived learning situations which simulate their 'real life' equivalent situation. Much in the manner in which a corporal in the army learns how to lead his section of seven soldiers, by doing so on field exercises conforming as closely and realistically as possible to the conditions likely to be encountered on the battlefield, the frontline leader from the business world is exposed during training to similarly simulated experiences. Though this could be called classroom learning it is not of the abstract or academic sort, it is *experiential*, and as such the lessons learned are not only immediately transferable to the real work situation, they can be reinforced with on-the-job *coaching*. This is a highly effective form of teaching.

Leadership then may be learned by the trainees undergoing a series of exercises. Probably one of the most effective programmes for inculcating frontline leadership is the series devised by Zenger–Miller and the Achieve Group of Toronto, Canada, in which trainees are taken through a structured sequence of simulated situations which may be applied to their real life situations as soon as they have been learned. This kind of training is especially effective if it is combined with the improving of communication skills. Since we have seen that Leadership *is* Communication, one of the most useful teaching systems for the improving of communication effectiveness is the series of activities developed by Dave Francis called *Unblocking Organizational Communication* (see Bibliography). Taken together these two training systems help towards the realization of Deming's point seven, to help people (and machines?) to do a better job. To help *machines* to do a better job calls for a different set of skills, namely those of statistical analysis which go under the general heading of Statistical Process Control (SPC).

The purpose of both leadership/communication training and the teaching of statistical methods is to enable us to make or do things right first time, every time. This begins to sound like sloganizing, and what does Deming tell us about *that*?

The Slogan Slingers

Slogans can be insidious in their effects. Like the one put up by the bosses of one particular organization who, wishing to proclaim to the world their recent conversion to the cause of Total Quality Management, erected a large sign on the factory wall right by the gate for all to see. In bold capitals it enunciated the managerial truism 'Quality starts here'. In equally bold capitals a wicked graffito added 'and here it stays'. That's one of the troubles with slogans, they often invite counter-comment. They can be seen as the counterfeit commitments of the non-committed. Because they can be bought in bulk they are able to spread like a rash across the face of the enterprise. You walk into the company's front office foyer to be informed 'Our business is Quality' by the slogan printed on the doormat; yet another pious proclamation of a hollow philosophy, more of the hypocrisy of our halfwit inheritance. Slogans are worse than waste of time, at best they are ineffective, words emptied of will. As Deming says, they are best eliminated. But what about 'zero

defects'? As a slogan it is as useless as the rest – as a goal it has
its merits, but the trouble is you will never know when you have
achieved it; as an operational standard it is worthless, impracti-
cable.

How about 'eliminate work standards that prescribe numerical quotas for the day'? We *live* by numbers. *We live or die* by
production figures. How can we even contemplate the elimination of numeric standards?

The Holy Cow

What you are about to read is iconoclastic. It is also seditious.

Iconoclasm is 'the breaking or destroying of images and
pictures set up as objects of veneration'. An iconoclast is 'one
who assails cherished beliefs or venerated institutions on the
ground that they are erroneous or pernicious.'

An icon is a statue looked upon as sacred.

The *Daily Output Figure* is an icon. Long ago in our manufacturing history this numeric value was turned into a sacred
statue, a holy cow. Holy cows are objects of cherished belief,
veneration and worship; they are set up and protected by the
established authorities, in the case of the Daily Output icon the
constituted authority is that of the Production Director, who
cherishes it.

Sedition is 'language inciting to rebellion against the constituted authority.' Therefore sedition can be a dangerous game to
play. Iconoclasm and sedition are frontal attacks upon the
status quo; the architects of the status quo do not like this, they
fight back. In our industries – both servicing and manufacturing – workers are employed to work. Supervisors (we use the
outdated term deliberately) are employed to oversee the
workers so as to keep them working. They work to achieve
output targets. If the targets are not met the workers are
always rebuked and often penalized. If the targets are reached
they are often thereafter elevated to become yet higher targets,
and sometimes the workers are rewarded. The aim of management is to keep the workers working harder and harder to
achieve ever higher targets. That's how to win in manufacturing
business. Or so goes the 'cherished belief', the holy cow of
Output Figures. But the truth?

A factory in which every worker is busy all of the time is a very inefficient operation. This iconoclastic opinion was expressed by Eli Goldratt, founder of Optimized Production Technology, creator of the Theory of Constraints, author of *The Goal*.

Try telling that to the Production Director. Try it even though his stocks of work-in-progress are cluttering the factory floor because his upstream production rates are overwhelming his downstream bottlenecks. Suggest, if you dare, that he cuts back on the upstream output by shutting machines down for part of the time. Try telling him that his quality problems have been solved and the 20 per cent output that used to disappear as junk now finds its way into the warehouse as prime stock, to add to the mega-piles of product already bursting the walls. He will tell you his problem isn't that he is making too much, it's that he needs a bigger warehouse.

This philosophy of production at any price, this slavish devotion to the god of big numbers, offends the three factors of Eli Goldratt's equation. 'Throughput' (the rate at which money is generated through sales) is down; 'Inventory' (money lying on the floor eating more money) is up; 'Operational Expense' (money used to turn Inventory into Throughput) is up. Therefore the business is running inefficiently. At this stage, no matter how excellent the quality of the product might be, the business is not generating the ROI it is capable of. So why is it being run this way? To nourish the voracious holy cow . . .

The Idiot's Tale: 8
Holy cow! Back to Bedlam. The place was a bloody shambles. But how the cowmen loved it.

Do you know what a 'shambles' is? It is a place of butchery and blood. A place to which animals are led to the slaughter. A place of death. The shambles of our manufacturing enterprises is the only place on earth, apart from the battlefield and the electric chair, where a sort of murder is legitimized. This is not an overstatement, in no way does it exaggerate the truth that the holy cow sometimes demands blood-sacrifice, and sometimes gets it. In Bedlam Big Boss was a Big Numbers man, therefore Little Boss was likewise. Big Numbers men are men who play the Big Numbers game. It can be a deadly game. Its rules are easily understood, even by the barely numerate. To play it all you need to appreciate is that the number 366 is bigger than the number 365. All that you need to believe is that the bigger the number the better, and biggest is best. The game is played with a very limited number of chosen numbers whose magnitude is deemed to be an all-embracing measure of total organizational performance – you will get the idea in a minute – and in Bedlam's case the Chosen Number was the *Output Figure*. Because belief held that the bigger this number became the better it would be, all company effort was geared towards maximizing it. A secondary, back up number was also chosen – Output per employee – to indicate how economically the Output Figure was being achieved; after all, they said, anybody can generate higher output by hiring more hands, but it takes good man-managers like us to do it with fewer, so we shall keep the headcount down to push the ratio up. Both these numbers were routinely communicated to Head Office, to show them how well Bedlam were husbanding the resources entrusted to them. This pair of complementary numbers were accorded total attention, and all other numbers were ignored. Inventory figures for quarantined work-in-progress were neglected; figures for output of dubious quality were banished by overlooking the 'dubious quality' aspect and sweeping them into 'output'; stock figures for rejected raw material were laid at the suppliers' doors as being not of our making. ROI was never even thought of.

Now there *are* times when the pursuit of big numbers is an entirely praiseworthy activity, so let us not throw any babies out with the bathwater. For example, during the running up of a manufacturing operation when productivity is moving in fits and starts, and the market is pulling for more, the pursuit of bigger output numbers makes obvious sense, it makes the *only* sense. If the output figure is lower than the business breakeven point you are not yet running a business: low figures are nothing more than promise without performance. Performance counts. High performance is rewarded. The output figure is a direct measure of the performance of the Production Manager. Bedlam's output figures were low, much lower than budget and lower than breakeven. So the Production Manager was clearly failing to perform. Low performance is punished.

There is an irresistible logic about this. Both Big and Little Boss loved what they termed 'logic'. 'Give me facts, give me logic,' snarled Little Boss as he looked at the disappointing output figures. The facts he had got – low numbers; the logic followed – it must be the fault of the Production Manager. He had the evidence, he arrived at the verdict – guilty as charged. All he had to do now was pass sentence. Never mind all those technical 'reasons' for the low output, they are just cop-outs, excuses; we cannot sell cop-outs, customers cannot use excuses.

The times bring forth the man. There is a certain type of managerial personality which is of nearly limitless ability in its own strictly limited field. Production management, in its narrowest sense of output driving, is one of these narrow fields. Bedlam badly needed such a manager, so Bedlam's time obligingly called one forth. It is as if such people lie dormant, awaiting destiny's call, in the way some Australian toads lie entombed for years in sun-baked mud to emerge triumphant when their hour arrives. Little Boss, collector of facts and lover of logic, appointed a new Production Manager. The first the deposed incumbent of the office heard about it was when he was introduced to his successor by Little Boss, who informed him, in front of the entire management team assembled for the purpose of providing an audience, that in view of his failure to perform (fact) he was to be demoted (logic) and as of now he would head into this new man and take it or leave it that's the way it is because logic is ruthless. Got that? Any questions? The condemned manager, shocked, numbed, publicly humiliated, gasping and twitching like a fish on a hook, realized he had zero options. He, poor fish, was gaffed, filleted and gutted and already pickling in the tart liquor of soured self-esteem. Swallowing his bitterness, realizing there was nothing else he could do, his face lost, he acquiesced, accepted the number two position; the one thing he could do he did not do yet, he saved that for later . . .

Little Boss congratulated himself on his own ruthlessness, his willingness to grasp nettles, his courage in handling unpleasant situations; after all, somebody has to call the shots. The episode left him with a warm inner glow of self-satisfaction. He forgot an ancient truth, that those who live by the sword perish by the sword, but this was yet to come.

The new Production Manager was an excellent choice, a man perfectly attuned to his times. There is an element of luck in all this; the production process, hitherto hiccupping along, had by now reached the state of being technically debugged. The months of painstaking work were about to pay off. What a happy conjunction of events, a process honed up and made ready for high output and a man of destiny to drive it. He was a man of near-limitless ability in his limited field. Managers of this personality type are a boon to the organization at this 'flowering' stage of its growth. Their devotion to driving the output figures even higher is total, obsessive, almost monomaniacal. Focused exclusively on productivity, their iron resolution steamrollers all obstacles as flat as a squashed cockroach; like Sisyphus rolling his boulder ever upward no matter how steep the gradient, their effort is unremitting and the output graph climbs remorselessly higher. Their *impatience* with setbacks is more the ferocious *patience* of a man searching for lice; they urgently find and eliminate them. Endurance is their workmate, even on the Sabbath they rest not.

Their minds closed to any distractions of friendship or family or novelty, their love is their work and their work is their love. They are achievers. But theirs is not an Achieve style of managerial culture. They are incapable of creating the spirit of the huntingband brotherhood. Lusting after personal power to satisfy some dark and secret inner craving they are able to function only in a Power-style of organizational culture, in which all power emanates from themselves. They are beyond the autocratic, they are the Zeus-like authoritarians. Their chosen style works well, for a while, and their personal effort is total and unremitting. They earn their achievement. How do they do it?

Here is one of their ways: Suppose you are a production manager and you have, say, seven production teams heading into you, and each of these teams is headed by a team manager. Every week you tot up the production score of each of the teams, and rank them in order of accomplishment, one to seven. Then you call a production meeting of the seven team managers. You refer to the ranking order, but only two positions are of any interest; the number one spot, whose manager is congratulated on his achievement (this is called 'recognition') and the number seven spot, whose manager is rebuked for his lacklustre performance (this is called 'motivation'). Somebody who knows a bit about number-juggling points out that the very fact of ranking seven different numbers in a descending sequence means that one of them will inevitably be seventh just as another will be first, so why take exception to the seventh? You will say 'because it is the worst'. True, so it is, but is it *significantly worse* than all the others? If it is not then the rebuke to the manager whose ranking is seventh is not justified. Before you rebuke him we had best do a statistical test of significance, which assumes his score is not truly worse than any of the other scores. This would be a chi (pronounced 'kye') squared test, carried out to determine whether the difference in the ranking scores could have occurred by mere chance or was the outcome of some special cause. If this test indicates that the results you are observing are due simply to Dame Fortune, then blame her and not the team manager. Your reply to this reasonable suggestion to take a logical look at the ranking data is dismissive; you talk about anybody being able to prove anything with statistics, and anyway it keeps them all on their toes (this is called 'giving a sense of purpose').

Bedlam's newest blue-eyed boy was certainly one of this breed, and world-class with it. Soon his output figures were soaring, way over budget. No longer was the market pulling against a shortage of supply; the tables had been turned, the market was glutted with overproduction. Saturation had been reached, and then surpassed. Like the Japanese wristwatch makers who have been so successful that they have run out of wrists, or the razor blade manufacturers who have shaved all the world's chins, Bedlam found themselves able to oversupply a finite market demand. The holy cow of the output figures had never been so well-fatted – what a splendid cowman the new manager had turned out to be. The superlative output figures were transmitted to Head Office. To prove how economically these were being achieved with such a low employee headcount, the Output-per-employee ratio was also published. The fact that this ratio was kept high by the simple expedient of making everybody work twelve-hour shifts instead of eight-hour stints was not revealed – the resulting overtime costs were buried in the accounts. This 'proved' what good man-managers they all were.

At this point commercial wisdom suggested that output should be trimmed back and levelled off until further sales outlets (difficult to find in a supersaturated market unless you resort to dumping and starting a pricewar) could be negotiated.

Trimmed back? Output? You cannot be serious!

But the warehouse is full. Then rent more warehouse space.

But we have quarantined raw material, quarantined work-in-progress, and quarantined output cluttering up the place. Then rent *more* warehouse space.

Stock goes into rented accommodation, up goes the inventory cost and up goes the operational expense.

But the rented warehouses are miles away. Then hire a haulage contractor to shift the stuff into them.

Paying mileage money – operational expense up again.

As the Eli Goldratt equation tell us, when inventory and operational expense are high, true throughput is low; so return on investment is *down.*

But the trouble with this special breed of obsessive production managers is that they are *unstoppable.* They are incapable of doing anything except going on and on. Though Bedlam were bloated with surplus output, paying rent for warehousing and charges for haulage, they kept on striving to break the output records already achieved. Bedlam's Production Manager, hailed as the saviour, was jealous of his accolade. Not for him the cutting back of any little bit of what he had spent months building up.

His standing in company esteem having been elevated in congruence with the output graph, he had by now acceded to the folk title of Mr Boss, thus joining the other two Bosses in the pantheon. To *reduce* output voluntarily might be to put this award at risk, so output was driven as if there was a famine in the marketplace.

This is the holy cow of Deming's 'numeric standards'. It is a voracious eater, whose appetite sometimes must be curbed. It begins by producing profits, and often ends by eating them. He recommends the elimination of numerical targets and standards, which to numbers-drunk managers is inconceivable. They assume him to be advocating the abandonment of all ambition and measurable accomplishment. He is not. When the advice of his points covered in this chapter is followed another paradox occurs – when the holy cow of Output is killed, the business does not suffer.

But this was Bedlam, the bloody shambles, the place of death. Whose death? Who was it that died? Within three months of being deposed the ex-Production Manager fell sick with a kind of wasting disease. His hair fell out, his countenance wizened. Perhaps this was a slower, Westernized version of what the Samurai calls *sepuku*, and the ordinary Japanese hara-kiri, which is what they do when they experience inconsolable loss of face. Having zero options on the day of his public degradation the humiliated manager exercised his last option – he gave up his life. A ritual suicide? A legitimized murder done slowly? Who will ever know?

Perhaps it was a blood-sacrifice to the holy cow of numeric standards.

It is more humane to do as Deming says, and kill numeric standards.

Summary

Part of the transformation of Western managerial style into the way of the New Order is a re-appraisal of the role of supervision. Supervision, in its old sense, is no longer acceptable because it no longer works. Neither do slogans, exhortations, buzz-words or trying to 'manage by objectives'. Management by control is evolving into *leadership* which is itself devolving towards the shop-floor.

The key to its success is improved communications. This can be taught, and therefore can be learned. This spells death to the holy cow of the ouptut for outputs sake figures. We are now dealing with iconoclasm and sedition – or with a healthy challenge to the status quo and to habitual thinking and belief. The new way of managing must first destroy the credibility of the old.

Bedlam plc were hooked on older thinking. Big numbers were the be-all and end-all. Aaron Godman had to alter this. Numeric standards sometimes result in localized optimization of business performance, an unbalanced operation, which must be corrected.

9 DEMING'S EIGHTH, NINTH, TWELFTH AND FOURTEENTH POINTS

8 Drive out fear, so that everyone may work effectively for the company.

9 Break down barriers between departments. People in research, design, sales, and production must work as a team to tackle usage and production problems that may be encountered with the product or service.

12 (a) Remove the barriers that rob the hourly worker of the right to pride of workmanship. The responsibility of supervisors must be changed from sheer numbers to quality.

12 (b) Remove the barriers that rob people in management and in engineering of their right to pride of workmanship. This means, *inter alia*, abolition of the annual or merit rating and of management by objective.

14 Put everybody in the company to work in teams to accomplish the transformation.

Again, it is convenient to examine several points together. Deming's eighth point, about driving out fear, is probably the most important of all his points.

Fear

Drive out fear! Whose fear? Fear of what? Fear of whom? *Is* there fear in the organization? *Should* there be fear in the organization? What does this point tell us about ourselves and the way we work together in organizations? Is Deming perhaps reminding us of something we would rather forget? What is this point really about?

It adjures us to drive out something whose very existence we are reluctant to acknowledge. It charges and entreats us solemnly, as if under the penalty of a curse, to get rid of something we would rather not admit to experiencing in our should-be-secure, oh-so-cosy, world of work – *fear*. This adjuration even states the 'curse' which will fall upon us if we fail to do as it says, 'so that everyone may work effectively for the company'. It warns us that if we fail to drive out fear we shall be cursed with an *in*effective workforce. This already sounds familiar. This curse of ineffectiveness has long been upon us, to a greater or lesser degree it has always been our norm. Is this because we have always used fear in our organizations – Far from driving it out we have invited it in and honed it to cutthroat sharpness, and deliberately used it? Perhaps we have always acted upon Napoleon Bonaparte's dictum 'the two mainsprings of human activity are self-interest and fear' as if his words were the only truth of human motivation. Carrots and sticks.

Carrots and sticks are not fit for encouraging anything with more dignity than a donkey. Yet we use them to 'motivate' human beings, the two edges of the sword called 'fear'.

Have *you* ever been made afraid? In your job, that is to say, not in any special circumstances like being almost drowned, or having a close shave with a speeding lorry whilst crossing a road – made afraid in your workaday life. Have you ever felt fear? Do you perhaps feel a little afraid right now? Have you ever set out to make anyone else feel afraid, a colleague, a rival, a subordinate? Surely though, fear is a perfectly natural emotion, an instinctive survival mechanism: how would it be possible to 'drive out' anything as deeply embedded in the human psyche as this? Fear has purpose. When confronted with danger fear offers us two options – fight or flight.

Exactly! Only the fearful take flight, and only the fearful attack. If the fearful cannot take flight because their escape route is blocked, then they will fight. The next time you believe anybody is 'attacking' you please remember, *only the fearful attack*. Oh, to be sure, they usually rationalize this attack by calling it self-defence. Those who attack nearly always adopt a stance of injured innocence, of affronted correctitude, of righteous retaliation. The perverse purpose of their attack is to exorcise their own fear, to drive it out of themselves by making it somebody else's. But this does not work, this cannot work, it cannot help but multiply fear by provoking retaliation. Fear remains within the attacker, but its tentacles have now been extended to include the attacked, who now, feeling fearful, counter-attacks. So fear feeds upon fear. It is a self-defeating trap.

There is a way of breaking this fearful trap, a way of driving out fear as Deming recommends (without advising us how to do it). A way so simple it seems miraculous, so powerful it is immediate, so proven it cannot fail. We shall be coming to it a little later; for the time being we shall look at another way of getting rid of fear, the way of knowledge.

The prospect of stepping out of the door of a high-flying aeroplane into thin and insubstantial air must be daunting and fearful even if you are wearing two parachutes. Parachutists tell us that it is the prospect of making the *second* jump which is most frightening of all. The very idea of falling from a great height is enough to bring on a giddy feeling of vertigo, so how is this perfectly natural fear overcome? With knowledge. The legend inscribed over the gate of the Royal Air Force parachute school reads 'Knowledge dispels fear'. Once fear has been driven out parachutists are encouraged enough to enable them to make any number of jumps. En-courage, meaning to inspire with confidence. So if knowledge is able to dispel the fear of those whose business is falling out of the skies, what about the earthlings like us? Through knowledge – training and education – might we meet Deming's Eighth Point and succeed in driving out fear? And will that result in people working more effectively for the company? Is Deming right? He is right, even though he says we should do it but omits to tell us how. It works whenever and wherever it is sincerely tried. This is attested to by the experience of those who have done it. But so far it seems too few are trying it. Knowledge does dispel fear, for those strong enough to test it out. But somehow not for long, and even then only here and there. Fear seems to run through the organization like the tendrils of couch-grass through a garden; deeply-embedded, impossible to eradicate, self-perpetuating and insidiously vigorous, ready to reinvade any weeded patch the moment your back is turned. Like original sin, it is there from the very beginning, you can find it right back there in the Contract of Employment itself. It is there in the 'jealous god' clause, did you notice it when you signed your employment contract? It's there all right, safely hidden in the thickets of the small print. It's the one which reads something like '. . . in the interests of the integrity of both the jobholder and the company the jobholder shall not take up any other gainful employment, nor be in receipt of any fees or emoluments or gratuities or other forms of income . . .'; it carries on in this pseudo-legal jargon before concluding '. . . without first obtaining the written permission of a Director of the Company . . .' or words to that effect. It sounds fair enough, and anyway, you want the job, so you sign the contract, exclusivity clause and all.

Now what is all this '. . . in the interests of the integrity . . .' bit about? Here we might have a manufacturing company pouring poisonous effluent into the river, gassing wildlife for miles downwind, laying workers off without a pang of conscience the moment trading turns down, wringing every last drop of reluctant sweat out of the labour which is left, and now there appears in the contract a sudden flush of morality with all this talk of *integrity*. In the interests of *your* integrity? Do you believe it? Has it not occurred to you that what this clause is really telling you is that they want you captive, dependent, vulnerable, submissive, docile? In a word, *fearful*. The ink of your signature to the contract is hardly dry yet and already the threat is stitched into the fabric of your job. Their opportunity to use fear as a weapon against you, by claiming sovereign rights over your *security*, is already institutionalized, the electrodes are in place, the switch to galvanize you can be thrown at will. They want you fearful, that way you can be all the more easily 'managed'.

Why? Because they too are fearful, and they want you in the same communal trap. In case you have not yet come across this kind of situation there follows a small episode, a scenario as they are fashionably known, just to show you how it works. By the way, this is *not* a trivial tale. It is entitled . . .

THE FEARFUL TRAP

Imagine yourself to be a young manager in a manufacturing company. It's your first managerial appointment. You know it's managerial because your name appears on the organization chart alongside the job title 'section manager', so it must be managerial. At last you are a *somebody*. You are pleased with your steady little job which you perform from the secure uterine confines of your shabby little office which has *your name* on its door. The door opens. In stalks your head of department, your boss. With one negligent push he sweeps aside the litter of paperwork to clear himself a perch on the corner of your tatty little desk. He flicks away some invisible specks of dust before risking the pinstripe of his trouser leg on the peeling plywood corner. Having gingerly settled on his roost, swinging his dangling leg like a slow pendulum, he deigns to open the conversation.

'How's things then?' (What things? you wonder.)

'Oh, so-so, you know.'

'Wife all right?' (They tell you to ask your subordinate that, when you go on that human relations course; always ask how the wife is, they say.)

'Fine, she's fine.'

'Fine,' he nods, 'fine.' Then he goes on, 'Pregnant, isn't she?' adding unnecessarily 'again.' (He believes himself to be a true representative of Rational Man with Average Family of one wife and one child. He thinks any more than one is carelessness and more than two is madness.)

'Yes, pregnant. Again. Four months pregnant.'

'Four months,' he muses, shaking his head with the wonderment of human fertility unbridled, 'and you've already got . . . three . . . is it three?'

'That's right. Three. Two boys and a girl.'

'And in another five months you'll have *four*.' He cocks his eyebrows, amazed by the folly of such fecundity and pleased with his arithmetical skills.

'That's right. Four.'

'Well, well, well. You'll soon be needing a bigger house,' he advises you; 'that means a bigger mortgage. Still, I suppose the little place you've got is quite cosy.'

'Yes, you could call it cosy.' Or you could call it a pokey little hole.

'You couldn't really afford a bigger mortgage though, could you?' he sadly commiserates with you and your straitened circumstances. 'Not on your salary, and what with *another* mouth to feed.'

'We'll manage somehow.'

'Oh yerse, manage, yerse, just about get by all right on
your existing salary, but I'll bet your wife and kids need
every penny of it, eh?' He leans forward conspiratorially
like an insurance salesman who has been to a seminar to
learn how to close that deal. 'But have you ever considered
what they would do if you were to *lose your job*?' He
straightens up, sits tall, his head higher than yours. He
stares solemnly at you down his nose, you look up to him.
'Have you ever *given thought* to what might happen to
your loved ones if you, their *breadwinner*, were suddenly to
find himself *on the dole*?'

Yes, you have given thought.

After another pause, to permit the full horror of this
possibility to cast a permanent shadow across your so-called
career path, he lopsidedly smiles his old-buddy smile. 'Still,
it's all hypothetical, isn't it? Purely hypothetical.'

No, it isn't. Neither pure nor hypothetical – it could
happen, and we all know it.

'A clever man like you wouldn't do anything silly like
losing his job now, would he?'

No, he wouldn't.

'Oh!' He is suddenly struck by an afterthought. 'While
I'm about it, I almost forgot to mention, it nearly slipped
my mind, the company has decided to have a bit of a
reorganization.'

Has it? First you've heard of it. You thought you were a
member of the company, but perhaps you are no longer
numbered among that happy throng.

'Yes, a bit of reorganization. It's like this, you see, Jack'
– he is now in his user-friendly mode – 'we have decided
that to have both you *and* Harry heading into me as two
section managers is a bit untidy, it would be tidier to merge
the sections into one, with just one section head – Harry –
reporting to me, and you reporting to Harry. To be fair to
Harry he *has* performed a bit better than you, so he is a
kind of natural choice, and anyway it will tidy things up, be
neater, cleaner, and . . . er . . . tidier.'

'You're demoting me, it's a downgrading.'

'No, no, no . . .' He seems to find the suggestion hurtful.
'Oh no, not downgrading, you will still be a manager, as
assistant manager, and you will still enjoy the same salary,
you will not suffer a drop of one penny. After all, you
simply could not afford to lose any money with all your
domestic commitments, so your salary will be preserved.
Until the next review anyway.'

He's got you. The fearful trap has sprung. You are in it;
vulnerable, dependent, without power. This is called management.

'Drive out fear,' says Deming, 'so that everyone may work effectively for the company.'

Be careful. Your boss has taken your family and, using them, has fashioned a weapon with which to attack you. Your immediate instinctive response is that of the wounded, cornered animal, to lash out. Hold it! You are too smart to do that.

What do you think of this situation? How do you feel about it? Here you are, innocent of any crime, humiliated for no reason apart from 'tidying up with a bit of reorganization'. It's not fair! You are angry. You've done nothing to deserve this demotion, but what can you do about it? A thought crosses your mind – 'don't get angry, get even'; then another one, the old Sicilian proverb 'Revenge is a dish best eaten cold'. The humiliation of it all tastes sour in the mouth, bitter as wormwood. The trouble is, you can too easily come to like the taste of it, grow over-fond of its corrosive taint, as you begin to plot your vengeance, and vengeance is *sweet* . . .

Stop! There is another way. A way so simple it seems miraculous, so powerful it is immediate, so proven it cannot fail. Remember, we spoke of it a while back?

What is it, this 'way'?

Let us consider this situation from a calmer standpoint . . .

There is always another way of looking at any situation.

So your boss 'attacked' you, using your own family and their dependence on you as his chosen weapon. He tried to make you afraid, he succeeded. Is this *true*?

Only those who are afraid choose to attack.

He attacked you, therefore he must have been afraid. Afraid of what? Perhaps of your expected reaction to what he had to tell you about the reorganization, so he attacked first, making a sort of pre-emptive retaliatory strike. This is clearly a kind of nonsense; it happens though, it happens a lot.

The strong do not attack.

Your boss attacked, therefore he was *not* strong. But surely, you will say, if the strong do not attack they will be assumed to be weak, then they will be attacked. So the strong must attack if only to demonstrate their strength. Otherwise they will be attacked. Will they? By whom? Not by the strong, they do not attack. Only the fearful attack, they are weak and therefore afraid of what looks like threat, so they attack it. But if the strong do *not* attack, then they are not perceived by the fearful as threatening, so the fearful *no longer feel afraid*, so they no longer choose to attack.

By this paradox may the fearful trap be broken.

So your best response to this 'attack' from your boss, young manager, goes roughly as follows:

While he is running through his preliminary rigmarole to remind you of your vulnerability ask yourself why he is afraid and therefore trying to make *you* afraid. Appreciate the fact that since he is weak, and that you are not weak unless you choose to be, you are in a position to *help him*. So you choose to be strong, and the strong do not attack. When he tells you that you will be downgraded and reporting to Harry ask him how he thinks you might best help Harry to achieve his – your boss's – objectives. But most important of all, during your dialogue with your boss, even while he is talking you must look at him in the weakness of his attack, and shout as mightily as you are able *but in complete silence* 'Jim, Eric, or Sid' (or whatever his name is) 'Jim, for this attack *I forgive you.*' Tell him again, in silent stridency, and *mean every word of it.* Tell him again.

You think this is metaphysical nonsense? You should try it some time. What the hell is all this rubbish about 'forgive', you say. Your scepticism is understandable. After all, he *demoted* you, didn't he? And did that trivial little act hurt anything except your vanity? That's what vanity is for, to hurt others and to be hurt itself. Did you not begin to plot revenge in the bitterness of your humiliation? Of course you did; vanity licks its lovely wounds and enjoys drinking from the dregs of the cup of bitterness.

So you forgive him. For doing what? For doing nothing except what your vanity imagined he had done.

What could be easier than that?

This is how you 'drive out fear', beginning with yourself, and then giving to other people. You cannot give it unless you have. You cannot have unless you give. You have by choosing to have, and by choosing to give. Strength.

But this is all too simple for you, my sophisticated and worldly-wise young friend. You know a better way, do you not?

Then why are you still afraid? Why do you still detest your adversary? You know the one we are talking about, oh yes, you know the one.

Just let your anger go. Stop nursing your wrath to keep it warm, it is taking up so much of your energy it leaves you too little to live on. It saps you of strength and drains you of joy.

But this advice amounts in your book to nothing more than mysticism.

So?

Mysticism takes over where your 'logic' leaves off. To what has your rationalist logic brought you? *To fear.* Mysticism, which Goethe called 'the scholastic of the heart, the dialectic of the feelings', opens the gateway to strengths within you, strengths you possibly never knew were yours. Do not question why it should be so, simply be grateful that it is. Try it, and see, *then* you will *know* (and knowledge dispels fear) with the knowledge of experience.

But in your intellectual need for a neat and tidy world in which everything is 'explained', in your desperation to understand something which is beyond all understanding, you are still hankering for some kind of 'proof' that this proposed mystical foolishness actually works. The only possible proof is the proving of it by personal experience, but to do this you will be required to renounce your vanity, just for a little while, just once in your transactions with your colleagues. This is asking too much for anybody steeped in the scientific method which believes that anything which cannot be measured cannot therefore exist. So let us see if it is possible to 'explain' the mechanics of mysticism . . .

NETWORKS OF CHOICE

Imagine that the universe in which you live is a net, like the ones used for catching fish but infinitely more complex and extending in all directions of space and time. You dwell upon one knot of this net, this is 'you' and 'now'. Your knot is connected by many filaments to many other knots, and ultimately through the totality of the net to all other knots. The filaments of the net are conductive, they transmit signals. The signals are either 'positive' or 'negative'; there are no neutral signals because you choose polarities. On the knot where 'you' live there is a switch. You are at liberty whenever you receive a signal and regardless of its polarity to flip the switch into either its positive or its negative mode. So no matter what the polarity of the signal you receive might seem to be, you can choose to send it on further into the network either as positive or as negative. You cannot avoid the act of choosing, you have no other choice than to select to send either positive or negative onward transmissions. In this network, this chaotic system, you cannot predict the outcomes of your choice; but there are 'laws' governing the operation of this system, laws of the sort being unravelled by the mathematics of chaos which is beginning to show us that even the state we call 'randomness' is lawful. But there seems to be one rule at work in this web of relationships which is of use to us, which is that the more positive signals you send the more you seem to receive, not on a miserly one-to-one basis, but as a vastly magnified return.

To those who are unaware of it all this is utter unintelligible nonsense, so there is no point in trying to know the unknowable and talking any further except to say that it is something whose effects can only be appreciated by those who have tried it, it can only be known by personal experience. To play a positive part in this web of unpredictability calls for a certain kind of courage – to be strong by being weak, to forgo vanity, to renounce ego. Those who have experienced it find it so simple it defies explanation, yet its outcomes can be little short of miraculous. This is the way, the Tao; but to know it you must try it for yourself. Try it, the next time you meet your adversary who sends you the negative signals. You know the one we are talking about – yes, that one. After all, what have you got to lose but your fear?

The Joshua Effect

Deming's points 9, 12a, and 12b which tell us to 'break down . . . remove . . . the barriers between . . . that rob . . .' are concerned with smashing walls and breaching barricades. This is Joshua at work in the Jericho demolition business, but *he* had seven priests bearing seven trumpets of rams' horns to follow the ark of the covenant as well as a multitude to do the shouting. What do we have? Well, we have the shouting, but far from breaking barriers down it is more the yelling of frightened men on the battlements hoping to keep the walls standing. Success in these points really depends upon prior success with Deming's Eighth Point, the removal of fear, because it is this that cements the walls together, the fear of those behind them that the walls might ever tumble into rubble and let 'the enemy' in. So what have we got, apart from the shouting? We have the universal

tools of Statistical Process Control and Total Quality Management, SPC being a kit of techniques for transforming data into
knowledge, and TQM being a managerial policy which ensures
that the knowledge is properly disseminated and used. How can
they be used in relation to these points? Through training and
education; but by using these activities as a *force for cohesion* in
a divided community. A community – the company – divided by
the barriers (walls) which separate one function from another
and one hierarchical stratum from its adjacent levels. This is
based on the dictum 'Those who learn together will earn
together', which sounds suspiciously like one of those slogans
which Deming abominates, but which encapsulates our point –
the mix of people attending quality awareness seminars. Each
'awareness seminar' should be attended by a small core group
chosen from a selected sector of the business, which includes
everybody in that sector regardless of rank, augmented by
other people from other functions. This is designed to stress,
almost subliminally, the universality, the *one-ness*, of the quality message. It emphasizes that quality is everybody's business and everybody's child, that *we are all in this together*. The
'walls' that Deming speaks of are cultural barriers, psychological defence works, thrown up by the fearful. Their demolition,
Jericho-like, can be achieved by this act of communal communication – the awareness seminar – through the sharing of the
educational experience. It is probably the best approach, short
of a fanfare from seven rams' horn trumpets, that there is to the
business of breaching barriers. Unless you can think of a better
one. If you can – then you will be able to turn it into an *objective*,
within that scheme of things known as 'management by objectives'. This would run counter to what Deming says in his point
12b, 'This means, *inter alia*, abolition of the annual or merit
rating and of management by objective'.

    Abolish management by objectives! What do you think of
that, then?

    Not much, one supposes, not much at all, especially if MBO is
tied in with annual assessments and pay awards. What is this
point really driving at? Is it the removal of objectives? This is
not what it says, it does *not* say 'This means, *inter alia*, abolition
of . . . objective.'

    By the way, if ever you doubted this book's core proposition

that quality is the new *religion*, you are now engaged in the kind
of disputation which is the hallmark of any religion, which is to
say the interpretation of holy writ, trying to elucidate what the
doctrine really says, what the guru is getting at. We have, willy-
nilly, become secular theologians, locked in esoteric debate.

But to return to Deming's recommendation to abolish MBO.
Surely it means simply what it says: not the abolition of
objectives, just the removal of that *managerial technique*
known as 'management by objective'. Once again it is the
removal of numeric standards, another killing of the old holy
cow of the Big Numbers game, but this time on behalf of the
manager. But to a generation of managers such as ours, weaned
onto a staple diet of MBO ever since we came off mother's milk,
any suggestion of giving up MBO is tantamount to a recipe for
anorexia, we *need* our MBO! Do we? Is there no other way?

SWEEPSTAKE ON THE DAILY RUN

Imagine yourself the helmsman on a sailing ship. Every
day, around about noon, your skipper appears on the poop
deck near your binnacle and takes a few sunshots through
his sextant, consults his nautical almanac, does a couple of
sums, chews his pencil, then puts a cross on the chart to
show you exactly where you are in terms of latitude and
longitude. 'How far are you hoping to take the ship in the
next twenty-four hours?' he asks you. You figure your
answer out – ship's making about eight knots, on an
easterly heading, wind coming from sou'west at about
fifteen knots, tidal current setting southward at three
knots . . . work out the vector and tell the skipper, 'Two
hundred nautical miles.' He takes his big dividers and his
distance rule, spaces their points two hundred miles apart
on the chart, one point in the present position as shown by
the pencilled cross, the other to its eastward; he pencils in a
little triangle around the eastern point of the dividers, this
is where you estimate your ship will be by this time
tomorrow. *This is your objective.* You are trailing a device
called a log astern of you, which measures your speed
through the water and totals up the distance you have run.
You set it to zero. You are a skilled helmsman, you've had
years of practice, your experienced eye rarely leaves the
compass in the binnacle in front of you, by regular slight
compensating adjustments of the wheel you hold the lub-
ber's line, which marks the ship's heading, in constant
alignment with the correct compass point. You have

allowed for wind and drift and calculated that your correct
course is two points southard of Cardinal Point East on the
compass's thirty-two point rose. You sail on, and on . . .

Soon be midday; up comes the skipper, scrutinizes his
turnip of a chronometer, takes his sextant and does his
sunshots, and goes through his daily navigational comput-
ing routine. He pencils his cross on the chart, whistles
through his teeth, shakes his head wearily, and calls you
over. 'Come and look at this,' he invites you. *This is your
assessment.* 'I thought you said you were going to get the
ship to *there.*' He stabs the little triangle of your estimated
target (your 'objective') that you agreed yesterday. 'But
you've brought us to here.' He indicates a pencilled cross
nowhere near the triangle. 'Miles too far off target and
miles too short.' You expostulate, 'But I said we'd cover
two hundred miles, and we did, it says so on the log, look.'
'Ah,' he shakes his head again, 'two hundred miles *through
the water*, not the same as two hundred miles *on the earth's
surface*, is it?'

'But we lost wind during the night, hardly a breath all
through the middle-watch,' you protest. 'Can't say as *I*
noticed it,' counters the skipper. Well, he wouldn't notice
it, would he, snoring in his hammock while you're up here
on watch. 'I can't honestly say that I'm all that happy with
your performance,' he advises you.

You have just been assessed by a managerial scheme
called 'management by objectives'.

'Be that as it may,' the skipper goes on, 'how many miles
do you reckon you can make in the *next* twenty-four
hours?'

You feel yourself boxed in by a numbers game. 'One
hundred and seventy five,' you venture.

'Hmph,' says the skipper, 'with *you* at the helm we shall
never make a landfall.'

It serves you right for being the helmsman on the Flying
Dutchman.

You'd have been better off on a clipper, they didn't go in
for MBO. Onboard the Cutty Sark your helm order would
have been 'Set course East by two South, crack on all the
sail she'll carry, haul taut and we'll be in China for tea.'
Goal-driven, no numeric standards, just mileposts on the
way.

Enough of salty yarns, back to Deming. He uses the term *inter
alia*, 'among other things', when he talks about the abolition of
MBO. What 'other things', one wonders. This is one of the
beauties of a good religion, it is a source of endless conjecture;

another good thing about it is that you are free to ignore it. So will you attempt the 'abolition of management by objective'? Of course you won't, and do you know *why* you won't? Because if you did you are afraid that you would not have much else to do. To you MBO is a manly thing, anything less is a 'mothering' approach.

But please note: MBO is to managers what work measurement and time and motion study are to the blue collar people – work divided, fragmented into little pieces served out with mistrust. An attempt to impose structures of certainty on that which is perceived as uncertain and therefore threatening – the future. An attempt to write the history of all that is yet to happen, a delusion of 'control'. This is why you love it, and why should you deprive yourself of your delusions as long as they keep you happy? Do they, though?

The Idiot's Tale: 9
The factory they gave to Mr Boss was not a very impressive structure whichever way they looked at it. Architecturally it followed the traditional form of a large rectangular box built of brick, with another longer, narrower box built along its frontage to form a run of offices. The conjunction of the office block and the factory wall formed a long empty corridor which served to baffle the factory's noise in order that the occupants of the offices might hear the dropping of pins. If its architectural form was disappointingly unimpressive, its financial form was even more so. It was, as it were, sagging into the quicksands of steady trading losses, and sinking deeper by the month. Mr Boss's job was to rescue it, to drag it out of the swamp and shore it up on a firmer foundation.

Mr Boss and I were standing at one end of the corridor. 'Do you know how wide this corridor is?' he asked. It was five and a bit plastic tiles wide, near enough six feet, but I knew this was the wrong answer, so I told him, 'Depends on the way you look at it.' 'This corridor,' he said, scowling down its long perspective, 'is a million miles wide.' He was a man who appreciated the value of hyperbole to emphasize a point. 'This *corridor*', he went on, 'is an impassable barrier between them in there', he waved a hand in the direction of the factory, 'and them on *this* side', he stabbed the air towards the offices, 'who live in all these little boxes, a *barrier to communication.*'

'That's exactly what Deming says in his ninth and twelfth points', I informed him.

'Deming? Who's he, where does he work?' So I gave him one of my infamous mini-lectures about Deming on Quality and Productivity. He appreciated it. 'Mmm, break down barriers,' he reflected. 'Right, we'll have some barriers broken down: starting this coming Saturday we'll demolish this wall, open the corridor onto the shopfloor, that'll do for a start.'

He acquired picks, crowbars, chisels, hammers, shovels, a wheelbarrow and leather gloves to protect tender palms; then he press-ganged all the managers and manageresses from the offices, telling them they were now part of a 'working group', and on a succession of Saturday mornings put them to work on the corridor wall, to remove it brick by brick. After a couple of months when this phase of the Jericho demolition job was completed he put them to more work within the office block itself, removing as many partition walls as was possible without bringing the roof down.

ᐟ 'Openness,' he declared, 'that's what management today is all about, accessibility, communication, living and working together as a team.'

Then he turned his attention to the factory itself. 'Have you *seen* it in there?' he asked with an air of incredulity. 'It's a shanty town, worse than Soweto. Come and have a look.'

Growth had been taking place within the factory, a kind of indigenous organic growth like the slow spread of some kind of architectural cancer developing unseen within the body of its host. Each 'cell' of future growth started its life as a chair, mated with a desk slung out of some office wallah's domain during a refurnishing spree. These two items of furniture were no sooner in conjunction than they began to grow a protective carapace around themselves, a shielding wall beginning with nothing more substantial than a sheet of packaging cardboard sellotaped into position. This embryonic wall would then extend itself by a process of accretion using materials salvaged from the environment, growing like coral, adding a spar or two here, a piece of plywood there, then a curtain of canvas and a sheet of acetate hung to keep the draughts out and pin-ups to paper the walls. There would then appear, within the privacy of this growing shell offspring; say a stool, or an electric kettle and some crockery, and some other things as well . . .

These structures proliferated oportunistically in whatever convenient niche was found to be free, so they now formed a scattered hamlet within the landscape of the factory. Like any hamlet some of the hovels specialized in the provision of certain services: in one you could buy sweets and cigarettes; another, run by an enterprising lady supervisor, provided buttered toast with waitress service on demand, made from the half dozen family-size loaves shipped in fresh ever morning by a van driver with a financial stake in the toaster. Another traded second-hand clothing and served as an ordering and delivery point for a catalogue club. In another of these shacks you could get your hair cut. Yet another did electrical repairs to broken household appliances, while someone else mended watches. You could purchase bird-tables fabricated from pallet-wood, or wrought iron gates cunningly constructed from the company's stock of mild steel bar. These were known as 'foreigners'. This was a closed order of mini-entrepreneurs responding to Adam Smith's 'unseen hand' of market forces, supplying demands. A culture of cardboard dwellings doing thriving business, clustering against the factory walls as thickly as the stalls crowding Jerusalem's Wailing Wall during the time of the Mandate.

'Shame to spoil it all in a way, I suppose,' Mr Boss remarked as he metaphorically sent in the bulldozers to clear away the rubbish of illegal private commerce, 'but it has to be done.' He seemed genuinely regretful of the necessity to stamp out this thriving economy. 'These people', he told his management team, 'are running little businesses within *your* big business. The trouble is *they* are running *theirs* better than *you* are running *yours*.'

He resolved to capitalize on this manifest talent and resourcefulness, to liberate it and direct it to the collective objectives of the company instead of to the splintered individualistic interests it had so far served so well. He embarked on a cultural change programme disguised as a quality improvement drive. In doing so he killed a whole flock of birds with one stone; he broke up the mini-Mafias of the factory floor; he raised productivity by improving process and product quality . . . he did all the things that Deming says are possible when the points of his doctrine are resolutely applied. This was the arrival of New Order management. It was a pleasure to work with such a battlefield general as this man turned out to be.

Summary

Probably the most important of Deming's Points is his eighth – 'Drive out fear'. Only then can all the barriers between functions and people be breached so that everyone in the organization is able to make his or her fullest contribution to the common wealth. Fear constrains, yet we must make free. So fear must be removed. Whose fear? Of what? How?

A practical guide to the removal of fear. The paradox of being strong by being weak, using the everlasting truth of loving your 'enemy' as a way of managing the situation. The folly of vengeance and the utter effectiveness of forgiveness. The way of the ego *v.* the Tao.

MBO is to managers what work measurement is to blue-collar workers – the institutionalization of official mistrust.

Harnessing *in*trapreneurial talent in Bedlam plc. Using new order managerial leadership to liberate and direct the capabilities of the workforce towards the organization's ends.

10 WHO NEEDS RELIGION?

So Quality may be regarded as a new secular religion. It has
many messiahs. We have looked at the doctrine of one of the
most eminent, and shall be briefly mentioning some of the
others. Like any religión it has its false prophets, its holy men,
its zealots, and its legions of peripatetic evangelists peddling
the bones of the saints and splinters from the true cross
promising salvation; they are called 'consultants'. But who
needs religion?

Religion? The very word carries too many overtones for
comfort, brings to mind too many unattractive images. Driz-
zling Sunday afternoons in the dusty dreariness of chapels
where the shrivelled congregations of the last of the devout
raise cracked voices in praise of a dead divinity. The guitar-
strumming rictus-grinning gaiety of the born-again, bright
Pentecostal eyes popping with the witness of personal revela-
tion on some private Damascus road. Droning ceremonial
observance of a doctrine over-decorated with form while drained
of all content. Doorstep missionaries with shining faces, hawk-
ing spiritual insurance policies, preaching that life gets better
when you're dead. Who needs *religion*? What is it for?

Everybody needs it; everybody always has needed it. Every
human society from the smallest tribe to the greatest nation has
developed its own religion. Its basic purpose is to tell its
devotees how to live at one with their world. Religion has its
roots in *ecological* imperative, it is about making the best use of
resources. For example, in these days of beef and butter mountains the dietary proscription that says you shall not have meat

and milk at the same meal makes no economic sense; to its founders, a pastoral people whose choice was *either* meat *or* milk but not both at once, it made excellent sense. So it became a 'law' of their religion. The purpose of religion is to make the most frugal use of resources; it is a discipline of thrift. *That* is Quality Management, a formalized and codified system of preventing the wastrel *mis*management of valuable resources, to ensure the safe perpetuation of the partnership between the resources and those who use them. It is a system of belief based on the knowledge of that which has been found to work; it is the distillation of experience, its purpose to save each generation from the necessity of having to find out for itself all over again knowledge which was already well-known. Quality management is such a creed.

Like any religion, Quality has one god but many sects, bewildering in their diversity. Quality used to be a simple-seeming, uncomplicated kind of a job – how is it that things have become so ravelled up that anybody wishing to do something about improving quality is faced today with a baffling array of options and advice?

The Invisible Ones

Quality, as has been stressed before in this text, is an *invisible input*. Invisible inputs are achieved by invisible people, and invisible people are lowly regarded. It has also been said before that quality used to be a despised discipline exercised by despised people, but nowadays it is a respected calling carried out by despised people. This is not said as a joke; if you think it is, and if you regard prevailing reward levels as an indication of professional esteem, look at job advertisements and rank them Pareto-fashion on a scale of salary by job function. Which function – finance, marketing, R and D, manufacturing, quality, personnel, distribution – gets the highest rewards and which the lowest?

There is nothing new in this. A few years ago quality management was seen by the business academies and prestigious consultancies as an 'artisan' activity, beneath their Olympian considerations. Nowadays the same institutions have become instant experts as supply responds to demand and the bandwagon riders jump on in their droves. This can make choice difficult for anybody wanting to improve his company's quality performance by making use of external education and training agencies. How to sort wheat from chaff, genuine from counterfeit.

Getting Some Help – Going out for it

A thousand shrines to the quality religion have lately been thrown up, temples of the cult proliferate, all offering 'courses', 'seminars', 'programmes', all teaching much the same thing. The range of what can be taught is not limitless; it is essentially confined to knowledge of the statistical tooling as mentioned in the early part of this text, but it is taught in much greater detail, and each school tends to emphasize its chosen aspect of the whole curriculum. There are numerous establishments offering first-rate instruction in the subject, so if you intend to send any of your people to these seats of learning how do you choose between them? There is a temptation to suppose that because some of these quality techniques are of American derivation it is best to go to America to learn about them. This is not necessary. In the UK there are many indigenous institutions as capable of teaching the subject as any in the world. There is a further temptation to seek guidance from the Japanese by going on a 'mission' to Japan. Save yourself the costs, stay home, lest when in Japan you hear the daunting words of Konosuke Matsushita:

> We are going to win and the industrial west is going to lose out: there's nothing much you can do about it, because the reasons for your failure are *within yourselves*.
> *Your firms are built on the Taylor model*; even worse, *so are your heads*. With your bosses doing the thinking while the workers wield the screwdrivers, you're convinced deep down that this is the right way to run a business.
> For you, the essence of management is getting the ideas out of the heads of the bosses into the hands of labour.
> *We are beyond the Taylor model*: business, we know, is now so complex and difficult, the survival of firms so hazardous in an environment increasingly unpredictable, competitive and fraught with danger, that their continued existence depends on the day-to-day mobilization of every ounce of intelligence.

For us, the core of management is precisely this art of mobilizing and pulling together the intellectual resources of all employees in the service of the firm. Because we have measured better than you the scope of the new technological and economic challenges, we know that the intelligence of a handful of technocrats, however brilliant and smart they may be, is no longer enough to take them up with a real chance of success.

Only by drawing on the combined brain power of all its employees can a firm face up to the turbulence and constraints of today's environment.

This is why our large companies give their employees three to four times more *training* than yours, this is why they foster within the firm such intensive *exchange* and *communication*; this is why they seek constantly everybody's *suggestions* and why the demand from the *educational system* increasing numbers of graduates as well as bright and well-educated generalists, because these people are the lifeblood of industry.

Your '*socially-minded bosses*', often full of good intentions, believe their duty is *to protect the people in their firms*. We, on the other hand, are realists and consider it our duty *to get our people to defend their firms* which will pay them back a hundredfold for their dedication. By doing this, we end up by being more 'social' than you.

This is what 'quality' is really about, improving the quality of people (by education and training), who will improve the quality of processes, which will then produce output of such superlative quality that it will not only 'meet the customers' needs', but will delight them with its quality. Or so goes the theory, so says the carefully constructed and maintained mythology. We shall be coming to myths and myth-making in a moment.

If you do decide to send people away to a reputable academy ask yourself about 'transfer-loss'. Transfer-loss is how much knowledge has spilled out of the bucket, the one they filled whilst on the course, by the time they've carried it back to base. These losses can be very high, a high as 60 or 70 per cent. This is not guesswork, but a measured value. Even so, if you wish only a few of your people to undergo training in quality management, then external courses are the appropriate way of doing it. If, however, you intend that more than three or four people be exposed to the educational process then it is probably cheaper on a cost-per-head basis to run the programme in-house, calling in an external teacher.

Getting Some Help – Bringing it in

There is a certain 'critical mass' consideration to be borne in mind when seeking to improve quality performance through training and education: this is the 'bonfire effect'. Have you noticed how, whenever you try to get an autumn bonfire burning in the garden, no matter how dry the rubbish or how combustible the hedge-trimmings, the fire refuses to go? So you splash more paraffin onto it – there is a brief roar of flame which quickly dies into a few wisps of oily smoke, and nothing hotter than sooty sticks to show for it. Too few twigs are burning, all the heat is being absorbed, the twigs lack enthusiasm and the licks of flame die down to an impotent smouldering. You reflect that being an arsonist must be a hard way of getting a living, and kick the cold heap to pieces to try over again. This time you build a core-blaze, feeding it until it suddenly starts producing, rather than absorbing, heat, its flaming tongues licking ravenously for more to burn. Now you have a bonfire: the more material you fling onto it the better it burns, green sprout trunks, whole sods of couch grass, it consumes anything.

It's the same with quality seminars in the organization anxious to improve its quality performance: expose too few people to it in the beginning and it will never reach critical mass. Besides, if you are bringing outsiders into the organization to lead the seminars, the more of your people attend each seminar the cheaper the per-head cost will be.

Should it be done 'top-down' cascade training, or 'bottom-up' starting at the shopfloor? *Both* these approaches *together* work best. The people at the top – the *enablers* – need to know what is going to be taught to their subordinates – the *doers* – in order that they will be able to manage the process of quality improvement. Sometimes there is a temptation to abdicate authority for the seminars in favour of the visiting consultant – in fact some consultants ask for it. Resist this temptation and retain sovereignty over the whole operation. The passing witch doctor is there as a guide, not as a usurper of your authority.

What qualities should you be looking for in the consultant? First and foremost – his *experience*. Has he actually *done* this thing he is teaching your people how to do? Done it for real, for a living, in the real world where real money changes hands and real things so often go wrong? Or is he going to teach a programme to a pre-ordained formula, with a set of colourful overheads and a big file which tells him which picture should be on the screen next and what words he should be speaking while that picture is showing? This is the myna bird method of teaching, sometimes the refuge of the inexperienced.

Does the consultant see you as a 'client' whose unique problems are to be addressed, or as a 'prospect' whose signature is wanted as quickly as possible on the next part of the contract? How hot is the hype, how much sausage is there under all that sizzle?

Quality management is *not* a magic wand, there are no fairy godmothers in this business, it is all about simple and straight-forward *work*, but it need not be boring. It can get a bit tedious to those who act as if it is all mathematical statistics, but it can be fun to those who realize it's about people and motivation as much as it is about mathematics, more so, in fact. This under-standing is the key to its success. To be effective quality education and training has to concern itself as much with motivation for change as with methods of doing things, and must seek to improve quality performance by altering company culture more towards the Achieve mode. It is about doing the things Mr Matsushita speaks of – going 'beyond the Taylor model' of work shattered by Scientific Management, D-realm work – and using quality to give more B-realm reward to those traditionally denied it in Western business philosophy. The application of the Deming doctrine is a powerful approach to bringing about this revolution in our thinking and hence in our actions.

This however – our inheritance of Taylorized work, with its implicit assumptions that people work only for money – is one of the most massive intellectual blocks standing in the way of progress towards higher quality performance. This reductionist view of life has soaked our thinking for such a long time in an acetic spirit of mistrust that it becomes difficult even to believe there might be a different, let alone a better, way. Three generations of Western managers have been born and brought up in the Taylorized way of work study, which shreds time into decimal places of a minute and lives by the numeric standards so antithetical to Deming's ideas. When some of these managers – sincere and honest men – find themselves in command of functions such as 'management development' there can be little hope of change for the better: their interest in the status quo is too entrenched, their cynical view (a cynic – one who knows the cost of everything and the value of nothing) of the irredeemable baseness of human nature distorts their perceptions. It is a bit like putting a Herod in charge of an orphanage and wondering why the boys don't grow.

There are other things which might divert us on our road to excellence; one of the most pernicious of these is our hankering to be looked upon by our fellows as a person of some consequence, an overweening desire to be 'recognized' as a worthy person, or, as it has been described in American managerial literature, a 'good human being'. This pressure on individuals to conform to some ideal model which they *think* other people expect them to aspire to leads towards a voluntary renunciation of individuality, to a willingness to be less of a real person and more of a socially approved cardboard cut-out. In this context of self-applied social pressure the true purpose of work – which is to use as many as possible of the talents we possess to try and make this world a slightly better place to live in – is obscured; work, instead of being an avenue to achievement, becomes a path to social approval. The consequences of this collective psychological climate can be devastating, witness the fate of a soldier and an empire at a place called Singapore . . .

THE SINGAPORE SYNDROME

On the morning of Sunday, 15th February 1942, Singapore stood as a bastion of British Imperialism in the Far East. Its military commander, General Arthur Percival, had been promoted during the years of peace between the ending of the Great War in 1918 and the outbreak of the Second World War in 1939. During this long sojourn in idleness the British military organization underwent the changes which befall *any* organization when times are too easy – it cloned. Untried by any external threat it became more and more introspective, more and more concerned with the preservation of the status quo. In this atmosphere of untested stagnation it strove to become more and more like itself; in this forcing ground of conformity originality was perceived as eccentricity, and discouraged. Dissent was seen as disobedience, and not tolerated. Resourceful-

ness degenerated into ritual, vision atrophied into short-
sightedness, status usurped stature. Conformity became
the cardinal virtue, and soon the clones were in command.
Arthur Percival, a man of his time, knew that Fortress
Singapore was safe. It was protected from landward
attack by the impassable jungles and rubber plantations of
the Malay peninsula – and being *im*passable it is axiomatic
that no army, least of all a Japanese army, could pass
them, and connected to the peninsula by an easily defended
causeway, though it was inconceivable that the causeway
would even need defending. Protected from seaward by
batteries of 14-inch guns capable of sending any hostile
fleet to the bottom, the prospect of this fortress ever
falling to an invader was utterly unthinkable. So Arthur
Percival, clone of the flabby years of complacency and
child of his cultural conditioning, was unable to think the
unthinkable. The thinking of the unthinkable was left to
the Japanese.

By mid-morning on Sunday, 15 February 1942, Victor
('Monty') Muncaster, Army No. 844130, Sergeant, Royal
Artillery, IX Indian Army Division, began to suspect that
he was about to be betrayed by his ultimate boss General
Arthur Percival. The invading Japanese army had already
achieved the unthinkable by advancing down the 'impassable' peninsula; they had then thrown an iron mesh over
the breach blown in the causeway to stem their attack, and
had landed on Singapore island on 8 December 1941.
Singapore was thirsty; all its drinking water was brought
to it through a water main running down the causeway,
which pipe had been shattered when the breach to stop the
Japanese had been blasted in the causeway, because somebody had forgotten it was there. Monty and his mates were
firing at the enemy over open gunsights, in close-quarter
street fighting, and under the onslaught the Japanese were
falling back. Then the rumour started going around, we are
going to capitulate! General Percival has had discussions
with the Governor General Sir Shenton Thomas, they
think there are too many unburied dead, danger of cholera,
no drinking water, too many civilian casualties . . . At
three o'clock in the afternoon of that fateful Sunday the
Union Jack over Fort Canning was lowered and the Rising
Sun hoisted in its place. The Unthinkable had happened.
General Percival, 'cool, correct, colourless', had surren-
dered his invincible arsenal, and with it a hundred thousand soldiers and an Empire on which the sun had not so
far dared to set. The batteries of 14-inch guns in their
seaward-pointing turrets had never been fired during the
siege of Singapore. The story was put about that they were

incapable by design of traversing more than a restricted seaward arc of fire and therefore could not be trained to landward where the attack was coming from. This was a fabrication, a lie. They were capable of being rotated through a full 360° circle of fire, but Percival, nicknamed 'the Goat', forgot, and the guns remained silent. Even a goat needs a scapegoat, it seems. Monty and the other soldiers of misfortune were paraded in front of the Raffles Hotel to contemplate a bleak future and watch the troopship *Empress of Asia* blazing in the bay as she unloaded her reinforcements straight into the POW cages, whence they were marched into the limbo of the camps.

At this point the myth of Japanese invincibility was brought into being. After all, ran the twisted logic, if a Japanese army of 30,000 can overrun an impregnable garrison of 100,000 then they *must* be invincible. Myths are immensely powerful, and you can use them once you know how.

Monty Muncaster was liberated by Mountbatten's XIVth Army on 20 August 1945. He was lucky. But the clones were still in command and they had not done with him yet. Having betrayed him once, they were to betray him a second time. An emaciated wreck weighing less than ninety pounds on his release, he was restored to modest 150-pound health by 1953, and being an Army-trained mechanical engineer had no trouble in finding a job in the booming heartlands of British industry. He began making motor bikes in Birmingham. To his bosses, captains of industry cloned up during the war production years of cost-plus accounting and the postwar consumer boom of easy sellers' markets, it was unthinkable that anybody might pose even the smallest challenge to their industrial supremacy, least of all a defeated little nation on the far side of the world which imported everything and exported nothing. Japan, an impoverished member of the Third World poverty league, with no resources other than its people, defeated in war and too poor to pay for even its own defence, *pose a threat*? Unthinkable, of no concern at all to Monty's civilian generals-in-command. *Their* concern was with the preservation of a lucrative status quo and the expression of antiquated social values. Their ambition was not to become better manufacturers of better motor bikes. It was axiomatic that British is Best so why bother to improve, their goal was personal gentrification. Their motivation, instead of residing within their industries, lay elsewhere; in a social system of tiered classes where at each level you could look upward with deference and downward with contempt. This was the canker gnawing at the roots

of British industry, this insidious pressure towards prissi-
fication and eventual gentrification in a structure glued
together out of graded social insult. An entire industrial
edifice was built on the crumbling brickwork of English
country-house attitudes and decaying values; bosses
whose concern should have been the stewardship of their
enterprises, whose vision might have been of a better
future, dreamed instead of nothing more substantial than
admission to a level where they might be invited to chase
after a pack of dogs pursuing a fox. The role-model was
wrong. Its remnants are still with us. It cost Monty, ex-
soldier, ex-POW, his job when once more his leaders-
without-leadership were unable to think the unthinkable,
and having betrayed him the first time into the limbo of the
lost they betrayed him a second time into the limbo of the
dole queue; they capitulated their industrial Singapore.

The myth of Japanese invincibility was resurrected. Again
the self-poisoning logic of the defeated came into play – the
Japanese have beaten us, we are the best, therefore they must
be invincible. The Japanese themselves do not say this; they are
realists not romantics; in any case they are too modest and too
polite to even hint it. *We* say it, as an *excuse*. Some of us do not
say it, some of us know of British companies whose performance
is so high that none could surpass it. But we are too modest, we
are afraid of excellence when it is ours, we prefer the game loser
and the gifted amateur, so we keep quiet about it, and sigh with
relief when we lose it and are able to relapse gratefully into the
familiar mediocrity of the second-rate. Excellence is ours whenever we want it, and the trouble is, we so rarely *demand* it.

These are the traps of the past, but their springs are rusting
and their teeth blunting with the corrosion of the years; soon
they will be no more, these mantraps made from the attitudes of
a vanished past. Times are changing, and aspirations with
them; recent research into managerial expectations reveals that
today's professional manager is more concerned with such
things as having a higher degree of personal control over his
working life, using to the full the knowledge and experience
gained over the years, expressing creativity in a culture of work,
enjoying a degree of variety in labour, than with courting the
esteem and approval of others or earning more than enough or
joining a group of peers in an endless meeting. Today's
manager, the evidence tells us, wants to be more of his or her

own man or woman and less of somebody else's, than a previous generation of manager was prepared to be. Today's manager works higher up the Maslovian ladder in terms of perceived reward, is consciously seeking B-realm work, has more to give to society and is determined to give it. This is splendid news to everybody except the last few dinosaurs of that dying species of manager who craved gentrification; these are professionals, but wideminded generalists with it. Men and women of their new time. The 'Age of the Thinking Manager' is dawning.

So it looks as if the Singapore syndrome, with all that went with it, is now little more than a tide of history that has rolled over us. It has done its work of demolishing some decayed institutions, though a few are left to us as hollow shells for the delectation of the diehard nostalgic. It has swept away some superannuated beliefs in our sovereign right to unquestioning supremacy, so, no longer authorized to dip our bread into other people's gravy, we are obliged to work to make our own. The old Imperial god is gone, today's god is the customer. The customer is invincible, and that is *not* a myth.

So much for the traps bequeathed to us by the past, what of the traps that beset the present-day road to excellence?

Traps and Snares

Probably the first and second of the traps waiting for the unwary one who wishes to improve company performance by improving quality management are – bafflement, followed by exclusivity.

BAFFLEMENT

So many messiahs! So many names! So many doctrines! Deming, Juran, Crosby, et al. So many teachers in so many schools of excellence: Oakland in Bradford, Dale in Manchester, Mortiboys in Nottingham, Owen in Bristol, and so on. So many consultancies each offering its own cocktail. Which to choose? Whom to go to? Where to go? Call a few of them in for a chat, to explore your needs and what you believe to be your needs. You never know, if you call enough of them in you might find yourself in the position of the man who called in all the vendors

of garden cultivating machines and asked each one for a practical demonstration – after a few demos his garden was dug over
and he was able to stall his purchase of a machine for another
year. This is the exploration and identification trap. Care is
called for to find help which meets your negotiated need. Be
specific in what you are ordering, but not as specific and
exclusive as the industrialist who asked consultants to improve
his productivity by 20 per cent, and *nothing else mattered*. They
did – they installed a payment-by-result system in the machine
shops, which paid the men bonus only while they were cutting
metal. So the men made swarf, and 30 per cent of all components
ended up under size; this was the institutionalized production of
junk from a reward system of too narrow a focus. Ask around,
make enquiries, be sceptical. Decide who you want, then take a
small bite of the cherry, see how it tastes before biting any more,
do not sign large sum contracts straight away. Buy one day at a
time if you can.

EXCLUSIVITY

'This one is all right therefore all others are all wrong' is the cry
of the zealot. And of the foolish. No *one* system of quality
philosophy and technique can possibly be all things to all people
at all times; one is stronger where another is weaker, so a
discriminating mix makes good sense. Retain autonomy, it's
your business, it would be folly to abdicate control of any part of
it to a stranger whose interest is essentially of a temporary
nature, although the stranger's passing presence should make a
permanent impact.

   These are two of the traps lurking this side of the threshold,
before the journey has even begun. Skirt these and there are
others . . .

Dale's Pitfalls

Barrie Dale and Peter Shaw, at the University of Manchester
Institute of Science and Technology (UMIST), have been teaching quality management for many years. They are also engaged
in research on behalf of the UK government into the effectiveness of quality management in its application to industry. Their

published research findings are of interest and use to any company embarking on the SPC/TQM route to higher performance. The failure of manufacturing organizations to avail themselves of the benefits of modern quality management can be summed up in three awful words – Ignorance, Apathy, Blindness. Consider the essence of their research.

First there are the difficulties of *introducing* Statistical Process Control (and modern quality management):

Lack of Knowledge and Expertise

This must be some sort of Rip van Winkle effect. After untold millions of words have been written, spoken, projected and generally scattered to the thirty-two winds of the world there are people *still unaware*, who *still* do not know? Apparently so, this falls under the umbrella of Ignorance, but knowledge dispels ignorance as well as fear, so the remedy is to hand.

Lack of Action from Senior Management

One can only suppose they are too busy doing other things, such as attending meetings. This belongs in the Apathy basket of managerial sins. It probably follows from the 'lack of knowledge' mentioned above; if you don't know what it is you are supposed to do you can hardly be blamed for not doing it, can you? And ignorance is usually doubly bound by the fact that the ignorant are ignorant of their ignorance.

Poor Understanding of Purpose of SPC

This is understandable. 'What's it for?' they ask, and nobody can tell them much more than that it's something they should be doing because 'our customers are asking us to do it'. Which is a good reason for doing it, but not the *best* reason. This is more Ignorance, in its familiar straitjacket of the doublebind. Its *purpose* is to *make money*, which is why *you* are *here*, is the answer.

Lack of Training of Operators

More Apathy, stemming no doubt from ignorance, the original sin in the religion of management. The time has come to take a bite of the apple from the tree of knowledge: its consequences can hardly be worse than those of ignorance which bedevil you at present. Lack of training! Can you really believe it? But research shows it to be the case, so the remedy is obvious.

Then come the difficulties encountered *after* trying to introduce it.

Which Process to Apply it to

More Ignorance; in any operation there are so many, many processes. The answers to this one are firstly: introduce it to that process which you think is hurting you most, and secondly: only do so if your discriminating judgement tells you there is a fair chance of achieving a little victory. Remember this . . .

Quality management is only rarely about making a 'quantum leap' in operational efficiency, it is not of the 'in one mighty bound we leapt into a golden future' sort. It is about all the thousands of little decisions made each day in the business. Each is like a little grain of sand, of itself seemingly trivial, but added to other grains it amounts to something. Quality is about making more of these little decisions right and fewer wrong. Get enough little grains into the 'Right' pan of the business balance and you tilt it in favour of success. Get more of yours into the 'Right' pan than your competitor does and you are winning. That's quality, in action.

Apply it sparingly to begin with. You cannot tackle the entire battlefield in one onslaught, it can only be done wave on wave, project on project, one little thing after another. There is nothing glamorous about it, it's about work, and there has never been anything glamorous about that – if there had been the upper classes would not have given it all away.

Which Chart Technique to Use

This is the neglected tools bit, which tool for which job, yet more Ignorance. Or back to doing your own car maintenance, and asking which spanner you should use, the answer depending on the size of the bolthead, and guidance being found from Haynes' maintenance manual. It's the same with quality: refer back to the quality tool-kit, say Oakland's *Statistical Process Control*, and pick the technique designed to do the job in hand. As has been mentioned before in this text, this new religion of quality, unlike all previous religions, does not absolve anybody from the unpleasant necessity of having to think things out for themselves.

Which Product or Process Characteristic to Measure

Yet more Ignorance. Measure whichever characteristic is thought to be important enough to measure. If you are selling bags of flour and weight is important then this is the characteristic which is to be measured and controlled. Or find out which one is offending you, and measure *that*. Only you can choose from all the many characteristics – diameters, pH values, tensile strengths, unit weights, surface hardness, telephone rings before answering, dwelltime of customers on stools in a fastfood restaurant, breaking strain of knitting yarn, and a million more characteristics – the ones which are of legitimate and potentially profitable interest.

General Resistance to Change

More Apathy. Say no more.

No 'Champion' to Carry the Torch

Yet more Apathy to an unbelievable extent. Can there possibly be *any* organization in the whole wide world of business which lacks that person with a bellyful of fire who wants, more than anything in the world, to change the world? Incredible though this may seem, research tells us that this is indeed the case. This is apathy to the point of no hope, optimism frozen to absolute zero. Any champion needs, as we have seen elsewhere in the text, a mentor and a protector to hold the dogs off. Perhaps where there is no apparent champion there is actually one in embryo awaiting the arrival of the other two members of the trinity.

There are other, little, pitfalls. Things like allowing control charts to degenerate into wallpaper; letting quality tasks atrophy into sterile ritual akin to incense-swinging. But knowing of these traps gives warning on how to avoid them. Avoiding them, in order to implement a vision of excellence which might be achieved by using quality management as a vehicle for cultural change, requires the exercise of personal power. How much do you think you have got? How much do you *want*? Ask, and it's yours. Don't you believe this? Then consider the nature of power, as described by Galbraith (see Bibliography).

Power and its Three Prongs – the Double Trident

What is power? This is not an idle question posed just for the sake of filling another page with homespun philosophy. Anybody who plans to make any changes to the work organization *needs* all the power he or she is able to muster; *no* power does *not* equal no change, having no power means that any changes which take place will do so because of somebody else's exercise of *their* power. The possession of power includes the ability to direct the course which change will take.

Power is the engine which drives vision, without the power to turn today's vision into tomorrow's reality vision is nothing more than impotent daydreaming, weak wishful thinking. Change-agents need power as well as vision.

When you think about it most of us seem to spend a lot of time in the organization seeking power of one sort or another for one reason or another. Maybe we seek power as a means of freeing ourselves from the power of others; this is as a good a reason as any. Perhaps we covet it in order that we may exercise it over others; this is as bad a reason as any, and anybody who feels this way about power would do better to stand for Parliament in the hope of election to being able to boss people about 'for their own good'. Some seek power in order to make changes to the organization, changes which they believe will be for the better; this is OK as long as their opinion of the value of the proposed change is shared by those who will be affected by it; there has to be a moral dimension to the way power is exercised and the outcome it is intended to attain. Power without morality leads directly to tyranny.

None of this has yet answered the question 'what is power?' Let us try again . . .

Power is the imposition of my will on your actions.

Or, in its collective expression:

Power is the imposition of the will of one group upon the actions of others.

Imposition? How is 'will' imposed on others? What is the *nature* of this thing called 'power'?

In the forerunning book to this one – *Right First Time* – power is seen to be of two kinds. There is that which is allocated to the individual by the organization itself, the delegated power called *authority* which is 'given'. Then there is the second sort which the individual cultivates for himself, which is not delegated power, but is 'taken' by the individual and called *influence*. This model of the dual nature of power served well enough at the time; it has been superseded by a more comprehensive model developed by John Kenneth Galbraith in his book *The Anatomy of Power* (see Bibliography). We shall take a brief synoptic look at the Galbraith model to see how it can help us make controlled and measurable changes to our organizations.

We have already, three times, met the triad in this text: that of Mind, Means and Materials; that of Technology, Management and Leadership; that of Champion, Mentor and Protector. It is as if three-ness is something special, as if the number 3 is a magic number, because here we have it again, but this time a *double*-triad in our study of the nature of power. Galbraith identifies three *styles* and three *instruments* of power:

STYLE 1: CONDIGN POWER

'Condign' means 'worthily deserved, merited, adequate', and is typified by the Mosaic Law of an eye for an eye and a tooth for a tooth. As such it is to do more with balanced punishment than with the imposition of will upon others; however, this is the term Galbraith chose to use. In our context the word 'coercive' – 'to constrain by authority resting upon the use of force' – might be a more appropriate name for this style. This is an expression of power which most of us are, thankfully, unlikely to meet in our working lives. The imposition of will on the actions of others through the use of force has long been outlawed in civilized societies; these days nobody feels the weight of the overseer's fist or the toe of the boss's boot. This is the dictator's way of driving, of enforcing submission through fear of physical assault – the worst assault the majority of us might experience these days is a verbal one. This condign or coercive style of the use of power could be rendered into the vernacular as *Arm-twisting*.

STYLE 2: COMPENSATORY POWER

This is the imposition of the will upon others through the medium of reward. Reward is granted by the overlord to those who submit and comply with his will; it is withdrawn from those who do not. This is a style which must be familiar to many of us; it is close to the old carrot and stick, except it's now more-carrot or less-carrot. This is the basic principle of the annual merit rating, based on performance appraisal in a management-by-objectives system, of the sort deplored by Deming. This is the economic thumbscrew employed to punish the non-compliant. It is the power-to-deprive vested in a managerial structure built after the manner of a feudal hierarchy on a 'more heads equals more points, more points equals more prizes' salary scheme. This is little more than a philosophy of *Palm-greasing*.

STYLE 3: CONDITIONING POWER

There are two sorts of conditioning; there is the direct conditioning based on immediate reward or punishment intended to reinforce behaviour which the conditioner approves of and to discourage non-approved behaviour. This is external to the individual being conditioned. Then there is the internalized conditioning designed to encourage the adoption of approved behaviour by the individual because he or she holds this behaviour to be desirable and appropriate. This is the double-hook of imposing the will by convincing the conditioned subject that it is his or her own. This is a most potent and pervasive style of power, present in any organization or society. This is *Mind-bending*.

Each of these three styles may be exercised through any, or all, of three instruments:

INSTRUMENT 1: PERSONALITY

This is the means by which the 'charismatic' leader bends others to his or her will, by the sheer force of a dominant personality. It may be employed in the context of any of the three styles. This instrument is individualistic.

INSTRUMENT 2: PROPERTY

As a source of power the possession of property, in the form of money, capital goods, etc., has an obvious connection with the use of the compensatory style of power. The enjoyment of high rank in the work organization is in effect to hold the title deeds to the property of those below – their jobs and their pay. 'Property' may be used as a form of bribery to secure the submission of others to the property-owner's will. This instrument is institutional.

INSTRUMENT 3: ORGANIZATION

An organization is a group of people submerging, more or less, their individuality within the group in order to further the purposes of the group because these are beyond the scope of the individual. So the individual submits to the 'will' of the group, which then imposes its will upon other groups. Organization is the most powerful of the instruments of power. This is the collective.

This is a very brief glimpse of the Galbraith model of Power. The reader is urged to pursue this enquiry further (see Bibliography) because an understanding of power is essential to its most profitable employment.

The reader is further urged to ponder which of these styles of power through which of these instruments is being directed towards himself or herself. Their influence is as omnipresent as the myriad of invisible waves pulsing through the ether. This is not to recommend paranoia as a way of life, it is a question of who is trying to get at whom, and why? Since you are unable to avoid it, you might as well learn to recognize it. After that you might consciously decide to use it. It is just another kit of tools.

It is interesting to relate these three styles and instruments of power both to the Harrison/Handy cultural model, and to the levels on the Maslovian ladder. Within which cultural contexts would the different aspects of power find best use? To which rungs on the Maslovian ladder are they connected? This is part of the diagnostic technique designed to lead to a deeper understanding of that complex entity called 'the company', in order that any resources deployed to alter the organization might achieve the maximum effect by identifying the 'leverage points' of change. Is it worth it? What is the payback?

The 'Message'

Payback? Many companies are only partly aware of their true losses due to ineffective quality. These are sometimes assumed to be big enough to be irritating but not big enough to be worth spending too much time and effort on. Such costs may be schematically represented by the whirlpool of waste as illustrated in Figure 10.1. Once this vortex is reversed, by implementing good quality practice, then it is transformed into the sort of spiral of profitability depicted in Figure 10.2.

Quality is about making money. About making more money by making fewer mistakes and by opening up more opportunities by satisfying more customers, better than they have ever been satisfied before, and better than they dared to expect. Quality which is so good, because so much skill has been applied to the attaining of it, right the way back up the supply chain, that it may be taken for granted; an invisible input by invisible people.

This is the secular religion of quality. Every religious tract should contain a few commandments, how about these:

1   Work Smarter, Not Harder.
2   Get Closer to the Customer.
3   Listen to the Voice of the Process.
4   Challenge the Status Quo.
5   Maximize Work that adds Value.
    Minimize Work that adds Cost.
6   Work is more than Getting a Living.
    It is Making a Better Life.
7   We do what We Are.

LOW QUALITY = LOW PRODUCTIVITY

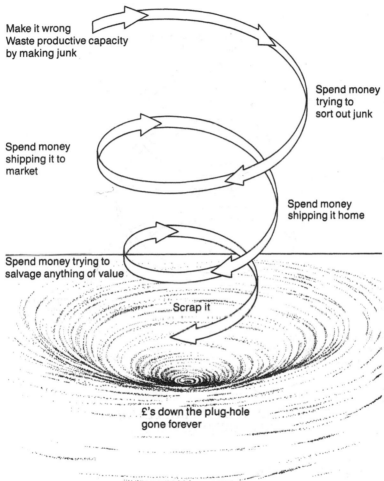

Make it wrong
Waste productive capacity
by making junk

Spend money
trying to
sort out junk

Spend money
shipping it to
market

Spend money
shipping it home

Spend money trying to
salvage anything of value

Scrap it

£'s down the plug-hole
gone forever

FIGURE 10.1 THE WHIRLPOOL OF WASTE
(QUALITY INCOMPETENCE)
(Quality costs for average company are between
25% and 40% of sales revenue

We Are whoever We Wish to Be.
and maybe . . .

8   When it stops Being Fun, Stop Doing it.

HIGH QUALITY = HIGH PRODUCTIVITY

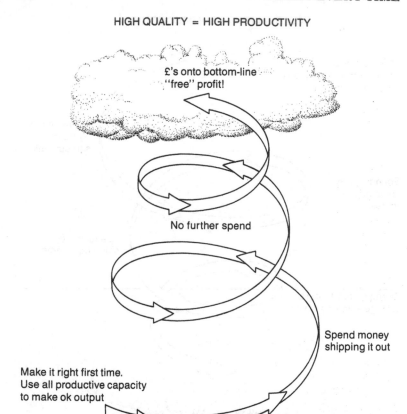

FIGURE 10.2   THE SPIRAL OF PROFITABILITY
(QUALITY COMPETENCE)

But these are only slogans.

Slogans they might be, and we know what Deming has to say about slogans. Even so, they are not entirely without value, they do tend to crystallize some of the elements of his doctrine. Like the Pole Star they can be used as navigational beacons, as points of focus for the mind. They are just words, but words are all that we – the writer and the reader – have to communicate with, and we are obliged to do the best we can in our Alice in Wonderland world wherein words mean whatever you want them to mean, so that whatever meaning they might assume is little more than that which the reader projects into them. Statements like 'Right First Time', or come to that, 'Right Every Time', or even worse 'Zero Defects': what do such terms signify? What is 'right'? How will you even know with complete certainty that you are actually achieving 'zero' defects? How is it possible to make any useful sense of such a non-operational definition? Are they anything more than pious catchphrases? As pious and hypocritical as another tired jargon-term from the lexicon of management – 'developing the full potential of the subordinate'? 'Full potential'? What is this supposed to mean? Are we expected to suppose that any society throughout history has *dared* to encourage or even permit the full development (whatever that might mean) of its citizens? Who then would hew the wood and draw the water? Those whose only innate potential was the mindless brawn of oxen? Any society – and that includes any corporation – in which everybody's 'potential' was developed to the full would burst apart under the force of expectations generated within itself. The purpose of the organization ·is to develop the potential of the jobholder only so far as it matches his abilities to the needs of the job, beyond that and the jobholder is turned over to the corporate psychologists for 'stress-counselling'; what kind of job is big enough to provide fulfilment for the infinity of potential which resides in all of us? Like the poet Walt Whitman, we are all vast, we all contain magnitudes, and it is an arrogant and demeaning assumption that a mere job of work, as experienced by the majority of our people, should be big enough to accommodate such an immensity of potential. But we assume it, as an act of conventional and collective belief. Which does not make it true.

This does not mean that we should deny anybody in their work the opportunities to make a greater contribution, by personal development, to the common weal. To do this we shall have to restructure the way we organize our work, to enlarge the scope of jobs to enable greater personal growth of the jobholders, we shall have to 'adopt the new philosophy' as Deming says, which is what the core of this book is about. The trouble is that anybody writing about the prodigious message of Deming's Doctrine finds himself in the same position as the man in the Greek fable who jumps after the frog. Each time the man makes a leap the frog does likewise, each successive bound halves the distance which separates them, but the man never manages to catch the frog. Close enough has to be good enough. So any reader wishing to get closer to the frog of this developing doctrine is urged, indeed earnestly recommended, to join the British Deming Association (UK readers, that is) whose address appears in the back of this book.

The reader will then be better equipped to add his or her voice to the dialectic of change, which must be made to happen if Western society is to work its way out of its present crisis. Bear in mind all the while that the greatest obstacles to change dwell within the organization itself – they are called 'colleagues', and the greatest obstacle of all is often to be found within *one's self. This* is where we must begin.

The Idiot's Tale: 10

Life's but . . . a tale
Told by an idiot,
Full of sound and fury,
Signifying . . .?

Signifying *what?* Things are no longer what they used to be, still, they never were. In response to their changing times, bosses and environment Bedlam have changed so much you would hardly recognize them any more. They have metamorphosed from one kind of madhouse into another. What was once a place of manic excitement and wild inefficiency has been lobotomized into a depressed Bedlam of boredom and dreary inefficiency. Still, that is, as they say, progress. If you can stop people having fun you must be getting somewhere.

Big Boss moved on, Little Boss did his sepuku act by dying in harness, other bosses came and went. The place was 'reorganized and prepared for building its future'. Reorganized in the sense that a Caribbean island is reorganized by the ferocious passage of a hurricane which uproots every tree and razes every shack, and so it is 'prepared for building its future'. This pleases those responsible for the reorganization; it affords them a sense of accomplishment whilst achieving nothing better than chaos and demoralization. That's what mediocrities in managerial office are for. Preferring conformity to competence, Form over Content, they build hierarchical pyramids, tiered tombs in which compliant zombies are interred according to arbitrary rank.

And what did Aaron Godman, full-time priest for ever after the Order of Melchizedek and part-time Quality bloke, do during this passing parade of temporary monarchs? Nothing much, on the face of it. Caused a little bloke, do during this passing parade of temporary monarchs? Nothing much, on the face of it. Caused a little rain to fall here and there, put back together a few of the broken things, spread a bit of gospel, sowed a few ideas. That's the trouble with visitors to industry such as we, no ambition. Our only legitimate goal is to make ourselves redundant once quality – the Invisible Input – is on the right track. But our biggest trouble is that we believe in the importance of certain eternal values which stand outside the framework of the organization; Justice, for instance. 'Do not expect or look for *justice* within the organization', proclaims the newest king as he practises his tyrant's injustice by summarily dismissing several of his staff whose only crime is that they happen to be on a payroll which he has sworn to trim.

So each of the monarchs has his day, and Justice stands aside and looks upon injustice. Soon it will be meted out to him, in his own coin. Until then . . . Ah, I must stop now, they have come to take me away, the two figures in cloaks of clinical white, difficult to know whether they are sons of light or psychiatric nurses, but all the same I have to leave you now. It was good to have been with you awhile in Bedlam, and I feel sure that we shall meet again. Until then . . . *think* about this comedy which is your portion, and *raise a smile*. This is the Age of the Thinking Manager, enjoy it, have fun . . . au revoir . . .

Summary

Quality has become the newest of the secular religions, with one god but many prophets. The prophets proliferate, messiahs are ten a penny. How to choose between them. Quality, the despised discipline of the West, a state of mind renounced by the East.

Where do you begin? How do you begin? By cascade or by sector training and education. Enlightened, and hence effective, quality leadership is inseparable from people and organizational development; it cannot be addressed independently of the culture in which it finds itself. Improvement in quality performance pre-supposes cultural change in the organization, to prepare the ground and render the climate conducive to growth.

The Singapore Syndrome and the making of a myth of invincibility. Thinking the unthinkable. Excellence is attainable. How and where to find help which will be cost-effective. Traps to be avoided.

The triple-prongs of Power, understanding it, cultivating it. Using it. Following the commandments of Quality.

SELECT BIBLIOGRAPHY

1 For anyone seeking a deeper understanding of the cultural environment:
   Charles Handy, *Gods of management*, Souvenir Press.

2 For those who wish to look further into the role of power, perhaps because they suspect they are being manipulated by 'them' and wish to counter-manipulate:
   J. K. Galbraith, *The anatomy of power*, Corgi.

3 For those who appreciate that communication is the essence of leadership and want more of both:
   Dave Francis, *Unblocking organizational communication*, Gower.

4 Want to know more about the Deming Doctrine? Then:
   W. Edwards Deming, *Out of the crisis*, Massachusetts Institute of Technology Center for Advanced Engineering Study.

5 Do you enjoy playing games? Try this as a guide:
   James P. Carse, *Finite and infinite games*, Penguin.

6 Want to know a bit more about who you are? This might help:
   Colin Wilson, *C. J. Jung, Lord of the underworld*, The Aquarian Press.

7 Want to know yet more about who you might become? This

will help:
A course in miracles, Author undisclosed, Arkana.

8 Background knowledge:
Frank Price, *Right first time*, Gower.

And where to learn even more about the Deming Doctrine:

The British Deming Association
2 Castle Street
Salisbury
Wiltshire   SP1 1BB
Phone 0722/412138

And where to discover yet more about a practical philosophy of applied Total Quality Management, or if you simply feel like entering into discourse about it:

Frank Price
Cerrig Llwyd
Waterfall Road
Dyserth
Clwyd
North Wales   LL18 6DB
Phone: 0745 570408

Finally, in case you are still not quite sure which is the takiti and which the golooma, the takiti is the spikey one and the golooma is the swirling curves.

INDEX

9 780367 456023